Medicinal and Aromatic Plants Used in Nepal, Tibet and Trans-Himalayan Region

Dr. Kamal K. Joshi and Prof. Sanu Devi Joshi

Bloomington, IN   Milton Keynes, UK

AuthorHouse™
1663 Liberty Drive, Suite 200
Bloomington, IN 47403
www.authorhouse.com
Phone: 1-800-839-8640

AuthorHouse™ *UK Ltd.*
500 Avebury Boulevard
Central Milton Keynes, MK9 2BE
www.authorhouse.co.uk
Phone: 08001974150

© 2006 Dr. Kamal K. Joshi and Prof. Sanu Devi Joshi. All rights reserved.

No part of this book may be reproduced, stored in a retrieval system, or transmitted by any means without the written permission of the author.

First published by AuthorHouse 5/16/2006

ISBN: 1-4208-6584-6 (sc)

Library of Congress Control Number: 2005906100

Printed in the United States of America
Bloomington, Indiana

This book is printed on acid-free paper.

Preface

Medicinal and aromatic plants are among the most studied groups of wild plant species of Nepal Himalayas. Yet information regarding their names, families, chromosome number, morphological descriptions, distributions, part(s) used, important biochemical constituent(s) and uses together in one place is not available. A comprehensive work with this information remained lacking since long at least of those Himalayan plants which are widely used or have potency for use for their medicinal and aromatic values. We have thus undertaken this task of providing information of those plants which are widely used in Nepal, Tibet and Trans-Himalayan region for their medicinal and aromatic uses. We hope this book will be useful to the university students and research workers studying economic botany, ethnobotany, medicinal plants and Ayurvedic medicine. It will also be useful to industrialists working on herbs, their extracts, herbal medicines and essential oils. Scholars and trekkers also may use this book for relevant information.

The authors are grateful to Prof. R. P. Choudhury, Dr. K. K. Shrestha, Mr. Suresh Ghimire, and Mr. Chitra Bania of Central Department of Botany, Tribhuvan University for providing literatures and valuable suggestions. Thanks are due to Ms Danielle Barker for her valuable help in arranging illustrations. Thanks are due to Ms. Nirmala Joshi of Department of Plant Resources of His Majesty's Government, and Mr. Prakash Manandhar for reviewing literatures and providing important information. We are grateful to Dr Saroj Joshi for suggesting us to publish this book through authorHouse and rendering help in various ways to make this publication a success.

Kathmandu
January 2006

K. K. Joshi
Sanu D. Joshi

ABBREVIATION

Alt.	Altitude
Am.	Amchi
C	Central
CITES	Convention on International Trade in Endangered Species of Wild Fauna and Flora Dolp. Nep. Dolpali Nepali
E	Eastern
Eng.	English
HMG	His Majesty's Government
N	North
NCS	Nepal Conservation Strategy
Nep.	Nepali
NEPAP	Nepal Environmental Policy and Action Plan
Npbh.	Nepal bhasha
S	South
Sans.	Sanskrit
subsp.	subspecies
var.	variety
W	Western
WHO	World Health Organization

Table of Contents

Introduction	xxv
Abies spectabilis (D.Don) Mirb.	1
Abrus precatorius L.	1
Abutilon indicum (L.) Sweet	2
Acacia catechu (L.f.) Willd.	3
Acacia nilotica (L.) Willd. ex Delile	4
Acacia rugata (Lam.) Voigt.	4
Achyranthes aspera L.	5
Achyranthes bidentata Blume	6
Aconitum bisma (Buch.-Ham) Rapaics	6
Aconitum ferox Wall. ex Ser.	7
Aconitum gammiei Stapf.	8
Aconitum heterophyllum Wall.	8
Aconitum laciniatum (Brühi) Stapf	9
Aconitum spicatum (Brühl) Stapf	9
Aconogonum rumicifolium (Royle ex Bab.) Hara	10
Acorus calamus L.	11

Aegle marmelos (L.) Correa	11
Agrimonia pilosa var. nepalensis (D.Don) Nakai	12
Albizia julibrissin Durazz. Var. julibrissin	13
Albizia lebbeck (L.) Benth	13
Allium carolinianum DC.	14
Allium fasciculatum Rendle	15
Allium wallichii Kunth	15
Alnus nepalensis D. Don	16
Aloe vera (L.) Burm. f.	17
Alstonia scholaris (L.) R. Br.	17
Amaranthus spinosus L.	18
Amomum subulatum Roxb.	19
Ampelopsis japonica var. mollis (Wall ex M. A. Lawson)	19
Anagallis arvensis L.	20
Androsace rotundifolia Hardw.	20
Apium graveolens L.	21
Arachis hypogaea L.	22
Arctium lappa L.	22
Arisaema intermedium Blume var. intermedium	23

Arisaema tortuosum (Wall.) Schott	23
Artemisia indica Willd.	24
Asparagus recemosus Willd.	25
Astilbe rivularis Buch.-Ham. ex D. Don	25
Azadirachta indica A. Juss.	26
Baliospermum montanum (Willd.) Müll. Arg.	27
Bauhinia purpurea L.	27
Bauhinia vahlii Wight & Arn.	28
Bauhinia variegata L.	28
Belamcanda chinensis (L.) Redoute	29
Berberis aristata DC. var. aristata	29
Berberis asiatica Roxb. ex DC.	30
Bergenia ciliata forma ligulata Yeo	31
Betula utilis D.Don	31
Boerrhavia diffusa L.	32
Bombax ceiba L.	32
Brachycorythis obcordata (Lindl.) Summerh.	33
Butea buteiformis (Voigt) Grierson	33
Butea monosperma (Lam.) Kuntze	34

Callicarpa macrophylla Vahl	35
Calotropis procera (Aiton) Dryand.	35
Caltha palustris var. himalensis (D.Don) Mukerjee	36
Canna edulis Ker-Gawl.	37
Cannabis sativa L.	37
Capparis spinosa L.	38
Carum carvi L.	39
Cassia fistula L.	39
Cassia tora L.	40
Cedrus deodara (Roxb. ex D. Don) G. Don	41
Celosia argentea L. var. argentea	41
Celtis australis L.	42
Centella asiatica (L.) Urb.	42
Chenopodium album L.	43
Cimicifuga foetida L.	44
Cinnamomum camphora (L.) J. Presl	44
Cinnamomum glaucescens (Nees) Hand.-Mazz.	45
Cinnamomum tamala (Buch.-Ham.) Nees & Eberm.	45
Cissampelos pareira Linn.	46

Citrus limon (L.) Burm. f.	47
Claviceps purpurea (Fr.) Tul	47
Clematis napaulensis DC.	48
Coelogyne cristata Lindl.	49
Cordyceps sinensis (Berk.) Sacc.	49
Coriaria nepalensis Wall.	50
Costus speciosus (J. König) Sm.	50
Cryptolepis buchananii Buch.-Ham.	51
Cryptolepis buchananii Schult.	52
Curculigo orchioides Gaertn.	52
Curcuma angustifolia Roxb.	53
Curcuma aromatica Salisb.	53
Cynodon dactylon (L.) Pers.	54
Cyperus rotundus L.	55
Dactylorhiza hatagirea (D. Don) Soó	55
Daphne bholua Buch.-Ham. ex D. Don	56
Datura metel L.	56
Datura stramonium L.	57
Delphinium himalayai Munz.	58

Dendrobium nobile Lindl.	58
Desmostachya bipinnata (L.) Stapf	59
Dichroa febrifuga Lour.	59
Dioscorea bulbifera L.	60
Dioscorea deltoidea Wall. ex Griseb.	61
Dioscorea prazeri Prain & Burkill	61
Diploknema butyracea (Roxb.) H. J. Lam	62
Dryopteris filix-mas (L.) Schott	62
Eclipta prostrata (L.) L.	63
Elaeocarpus sphaericus (Gaertn.) K. Schum.	64
Embelia tsjeriam-cottam (Roem. & Schult.) A. DC.	64
Entada phaseoloides (L.) Merr.	65
Ephedra girardiana Wall. ex Stapf.	65
Euphorbia hirta L.	66
Ficus racemosa L.	67
Flemingia strobilifera (L.) W. T. Aiton	67
Foeniculum vulgare Mill.	68
Fritillaria cirrhosa D. Don	69
Fumaria indica (Hausskn.) Pugsley	69

Gardenia jasminoides J. Ellis	70
Gaultheria fragrantissima Wall.	70
Geranium nepalense Sweet	71
Gloriosa superba L.	72
Gmelina arborea Roxb.	72
Hedera nepalensis K. Koch.	73
Hedychium spicatum Smith	73
Hemerocallis fulva (L.) L.	74
Heracleum nepalense D. Don	75
Hippophae tibetana Schltdl.	75
Holarrhena pubescens (Buch.-Ham.) Wall. ex G. Don	76
Hygrophila auriculata (Schumach.) Heine	77
Hyoscyamus niger var. agrestis (Kit.) Beck.	77
Ichnocarpus frutescens (L.) R. Br.	78
Impatiens balsamina L.	78
Inula cappa (Buch.-Ham. Ex D. Don) DC.	79
Inula racemosa Hook. f.	79
Ipomoea nil (L.) Roth.	80
Iris decora Wall.	81

Juglans regia var. kamaonia C. DC.	81
Juniperus indica Bertol.	82
Juniperus recurva Buch.-Ham. ex D. Don	83
Juniperus sibirica Burgsd.	83
Justica adhatoda L.	84
Leonurus japonicus Houtt.	85
Lepidium apetalum Willd.	85
Leucas cephalotes (Roth) Spreng.	86
Ligustrum nepalense Wall.	87
Lilium nepalense D. Don	87
Lindera neesiana (Wall. ex Nees) Kurz	88
Litsea cubeba (Lour.) Pers.	89
Lobelia chinensis Lour.	89
Lobelia pyramidalis Wall.	90
Lonicera japonica Thunb.	90
Lycium barbarum L.	91
Lycopodium clavatum Linn.	92
Macrotyloma uniflorum (Lam.) Verdc.	93
Maesa chisia Buch.-Ham ex D. Don	93

Maharanga emodi (Wall.) A. DC.	94
Mahonia napaulensis DC.	94
Mallotus philippinensis (Lam.) Müll.	95
Malva vercillata L.	96
Mangifera indica L.	96
Megacarpaea polyandra Benth.	97
Melia azedarach L.	98
Mentha spicata L.	98
Mesua ferrea L.	99
Michelia champaca L.	100
Mimosa pudica L.	101
Momordica charantia L.	101
Morchella esculenta (Linn.) Pers.	102
Moringa oleifers Lam.	102
Murraya koenigii (L.) Spreng.	103
Myrica esculenta Buch.-Ham. ex D. Don	104
Nardostachys grandiflora DC.	105
Neopicrorhiza scrophularifolia (Wall. ex Benth.) Hemsl.	105
Nyctanthus arbor-tristis L.	106

Nymphaea stellata Willd.	107
Ocimum tenuiflorum L.	107
Operculina turpethum (L.) Silva Manso	108
Oroxylum indicum (L.) Kurz	108
Osbeckia nepalensis Hook.	109
Osyris wightiana Wall. ex Wight	110
Otochillus porrectus Lindl.	110
Oxalis corniculata L.	111
Paederia foetida L.	111
Panax pseudo-ginseng Wall.	112
Pandanus nepalensis Kurz	112
Papaver sominiferum L.	113
Paris polyphylla subsp. marmorata (Stearn) H. Hara	114
Parmelia cirrhata Fr.	115
Parmelia nepalensis (Tayl.) Hook.	115
Parmelia nilgherrensis Nyl.	116
Parmelia tinctorium Nyl.	116
Parnassia nubicola Wall.	117
Perilla frutescens (L.) Britton.	117

Pholidota articulata Lindl.	118
Phyllanthus amarus Schumacher & Thonn.	119
Phyllanthus emblica L.	119
Phyllanthus urinaria L.	120
Phytolacca acinosa Roxb.	121
Piper longum L.	121
Pistacea chinensis Subsp. integerrima J. L. Stewart Rech. F.	122
Plectranthus mollis (Aiton) Spreng.	123
Plumbago zeylanica L.	123
Podophyllum hexandrum Royle	124
Polypodium vulgare L.	125
Potentilla josephiana H. Ikeda & H. Ohba	125
Prunella vulgaris L.	126
Prunus armeniaca L.	126
Prunus creasoides D. Don	127
Prunus persica (L.) Batsch.	127
Pueraria tuberosa (Roxb. ex Willd.) DC.	128
Punica granetum L.	129
Quisqualis indica L.	129

Ramalina Ach. sps.	130
Rauvolfia serpentina (L.) Benth. ex Kurz.	131
Rhamnus napalensis (Wall.) M.A. Lawson	131
Rheum australe D. Don	132
Rheum nobile Hook. f. & Thomson	132
Rhododendron anthopogon D. Don var. anthopogan	133
Rhododendron arboreum Sm.	133
Rhododendron lepidotum Wall. ex G.Don	134
Rhynchostylis retusa (L.) Blume	135
Ricinus communis L.	135
Rosa laevigata Michx.	136
Rubia manjith Roxb. ex Fleming	137
Saccharum spontaneum L.	138
Sapindus mukorossi Gaertn.	138
Saraca asoca (Roxb.) W.J. de Wilde	139
Satyrium nepalense D. Don	140
Saurauia napaulensis DC.	140
Schima wallichii (DC.) Korth.	141
Scutellaria barbata D. Don	141

Selinum wallichianum (DC.) Raizada & Saxena   142

Semecarpus anacardium L.   143

Sesamum orientale L.   143

Shorea robusta Gaertn.   144

Sida cordifolia L.   145

Sida rhombifolia L.   145

Sigesbeckia orientalis L.   146

Smilax aspera L.   147

Solanum anguivi Lam.   147

Solanum virginianum Dunal   148

Swertia alata (Royle ex D. Don) C. B. Clarke   148

Swertia angustifolia Buch.-Ham. ex D. Don var. angustifolia   149

Swertia bimaculata (Siebold & Zucc.) C.B.Clarke   149

Swertia chirayita (Roxb. ex Fleming) H. Karst.   150

Swertia ciliata (D. Don ex G. Don) B. L. Burtt.   151

Swertia multicaulis D. Don   151

Swertia paniculata Wall.   152

Symplocos paniculata (Thunb.) Miq.   152

Syzygium cumini (L.) Skeels   153

Tabernaemontana divaricata (L.) R. Br. ex Roem. & Schult.	153
Tagetes erecta L.	154
Taraxacum officinale F. H. Wigg.	155
Taxus wallichiana Zucc.	155
Terminalia bellirica (Gaertn.) Roxb.	156
Terminalia chebula Retz.	157
Tinospora sinensis (Lour.) Merr.	157
Trachyspermum ammi (L.) Sprague	158
Tribulus terrestris L.	159
Trichosanthes tricuspidata Lour.	159
Typha angustifolia L.	160
Uraria picta (Jacq.) Desv. ex DC.	161
Urtica dioica L.	161
Usnea thomsonii Stirt.	162
Valeriana jatamansii Jones	162
Vanda cristata Lindl.	163
Viscum album L.	164
Viscum articulatum Burm. f. var. articulatum	164
Vitex negundo L. var. negundo	165

Woodfordia fructicosa (L.) Kurz.	166
Wrightia arborea (Dennst.) Mabb.	166
Xeromphis spinosa (Thunb.) Keay	167
Zanthoxylum armatum DC.	167
Ziziphus mauritiana Lam.	168
GLOSSARY OF MEDICAL TERMS	217
REFERENCES	231
Index	233

INTRODUCTION

Bi-directional development in the field of medical science has reached to its summit for their promising effects man has quest for centuries and thus earned utmost popularity throughout the world. One of these two directions has been led by gene manipulation or using gene-related technology or genetic engineering and the other by preventing from or curing the diseases by the use of medicinal plants.

Recognition of values of medicinal plants remained confined to the developing countries for long but recently its popularity has been spread so widely that in some cases it is preferred to chemical control of diseases equally in developed countries. WHO (1989) identifies four main reasons for this widespread acceptance. They are:

1. Medicinal plants have been in use for untold centuries and have proved reliable and effective in treating and preventing disease.

2. Most species of medicinal plants are not toxic and therefore give rise to few, if any, side effects; even when some adverse effects do occur, they are much less serious than those caused by chemically synthesized medicines.

3. People living in rural and mountainous areas have easy access to local medicinal plants, so that their use in preventing and controlling disease costs much less than if Western medicine were used and is thus economically beneficial to developing countries.

4. Medicinal plants are an important source of practical and inexpensive new drugs for people throughout the world.

Himalayas are considered the big storehouses of enormous important plants and Nepal Himalayas representing the central Himalaya provide shelters to a large number of species distributed from few meters to around 5000 m above sea level. Great range of bioclimatic variation from tropical to the alpine zones brings richness in biological diversity in Nepal.

Nepal occupies a land area of 147,181 sq km between the latitudes of 26°22'N and 30°27'N and longitudes of 80°04'E and 88°12'E at the central third of the massive Himalayan chain 2500 km long. In more or less rectangular shape, Nepal stretches lengthwise about 885 km east-west and has an average width of 193 km north-south. The altitudinal concepts deriving from studies made in the European Alps has always been applied in describing vegetation of Nepal.

Schweinfurth (1957), Stearn (1960), Banerji (1963), classified vegetation of Nepal based on its bioclimatic variations both vertical and horizontal. Accordingly, Nepal is divided into three major regions: west (Long. 80°40'-83°0' E), central (Long. 83°0'-86°30' E) and east (Long. 80°30'-88°12' E).

Stainton (1972) identifies six main geographical divisions *viz.,* (1) Terai, Bhaber, Dun valleys and outer foothills; (2) the midland areas and southern sides of the main Himalayan ranges; (3) the Humla-Jumla area in the north-west; (4) dry river valleys; (5) inner valleys; and (6) the arid zone resembling more or less the Tibetan plateau in its character. Dobremez (1972) divides Nepal into six bioclimatic zones or belts with 11 sub-zones.

Customarily, the vegetation of Nepal is described in relation to its bioclimatic zones. Altitudes, latitudes, rainfall pattern, aspects of geology and human intervention are considered important environmental factors affecting the vegetation. Human impact and livestock pressures rapidly transformed vegetation of Nepal Himalayas. Forests of Terai, Bhabar and Dun valleys have largely been depleted for human settlements and agriculture purposes. Forests in midland mountains and valleys are also under high human pressure. Unfortunately, even the subalpine and alpine forests do not remain safe due to gradual increase in grazing and tourism. However, implementation of community forestry program aiming at helping conserve the natural habitats as well as restore the depleted forests by afforestation started showing promising results to a considerable extent in midhills. Still afforestation always remained beside the mark in copying the natural habitats. In fact, the afforestation programs never attempted copying them to the extent possible. Instead, all of these programs often confined in planting plants of timber values or those mostly used by respective local communities. The result is the continued genetic erosion of Himalayan plants even in the midhills where apparent improvement in restoration of depleted forest is recorded. It thus leaves the decision-makers and the users the choice between complete arrest of further depletion of natural habitats by total conservation of virgin forests and continued genetic erosion of Himalayan plants.

Analysis on the distribution of flowering plants and gymnosperms in Nepal reveal that 246 species out of about 5,188 species are endemic to Nepal. IUCN (The World Conservation Union), Nepal categorized their conservation status and listed 101 species under endangered and eight under feared extinct groups (Shrestha 1999). Most of these endangered species are the plants of medicinal and aromatic values, which are either confined in small area(s) and/or are collected recklessly from the nature for illegitimate trade.

Documented records show that there are 571 species of medicinal plants in Nepal (Anonymous 1993). The number may be several hundred higher than this, at least to around 700 species. Malla and Shakya (1984-85) enumerated 510 species of medicinal and aromatic plants from Nepal. These plants together with all indigenous species, wild or cultivated, in Nepal can be enlisted in the genetic heritage of Nepal Himalayas.

The signatories of World Heritage Convention (Paris 1972) commit under Article 4 to identify, protect, conserve, present, and transmit to future generations the country's cultural and natural heritage to the best of its ability. Under Article 5, the signatories further commit to:

- adopt a general policy which aims to give the cultural and natural heritage a function in the life of the community and to integrate the protection of that heritage into a comprehensive planning program;

- set up within its territories, where such services do not exist, one or more services for the protection, conservation, and presentation of the cultural and natural heritage with an appropriate staff possessing the means to discharge their functions;

- develop scientific and technical studies and research; and to work out such operating methods as will make the state capable of counteracting the dangers that threaten its cultural and natural heritage;

- take the appropriate legal, scientific, technical, administrative, and financial measures necessary for the identification, protection, conservation, presentation, and rehabilitation of this heritage; and

- foster the establishment or development of national or regional centers for training in the protection, conservation, and presentation of the cultural and the natural heritage and to encourage scientific research in this field. (See Shrestha 1999).

- Nepal as a signatory of World Heritage Convention since 1978, shares international obligation of protecting world heritage sites. Nepal's eight cultural heritage sites and two natural heritage sites are inscribed in the World Heritage List.

Other international conventions related to the conservation of genetic heritage to which Nepal is a party are:

Convention on International Trade in Endangered Species of Wild Fauna and Flora (CITES 1973)

- Convention on Biological Diversity (1992) and
- Wetlands Convention (Ramsar 1971)

Various policy and legislative measures have been taken in Nepal for the conservation and protection of its cultural and natural heritage. The National Conservation strategy for Nepal (NCS 1988), the Nepal Environmental Policy and Action Plan (NEPAP), the Environment and Resource Conservation chapter of the Ninth Plan (1998-2002) are worth mentioning among the policy measures taken by Nepal. The legislative measures include:
- the Ancient Monuments Protection Act, 1956;
- the Plant Protection Act, 1972;
- the National Parks and Wildlife Conservation Act, 1973;
- the King Mahendra Trust for Nature Conservation Act, 1982;
- the Town Development Act, 1985;
- the Forest Act, 1993; and
- the Environment Protection Act, 1996.

Eight plants or group of plants of Nepal are included in CITES appendix II *(viz., Ceropegia* sp., members of the family Cyatheaceae, Cycadaceae, and Orchidaceae, *Podophyllum hexandrum, Rauvolfia serpentina, Dioscorea deltoidea* and *Taxus wallichiana)* and seven species in appendix III *(viz., Cycas pectinata, Gnetum montanum, Talauma hodgsonii, Meconopsis regia, Podocarpus nerifolius* and *Tetracentron sinense).*

The Forest Acts (1993) of His Majesty's Government (HMG) bans *Cordyceps sinensis* (Yarsa Gomba) and *Dactylorhiza hatagirea* (Panch aunle) for collection, use, sale, distribution, transportation, and export. The same Act bans *Nardostachys grandiflora* (Jatamansi), *Rauvolfia serpentina* (Sarpagandha), *Cinnamomum glaucescens* (Sugandhakokila), *Valeriana wallichii* (Sugandhabala), Lichen (Jhyau), *Taxus wallichiana* (Talispatra or Lodh Salla) for export outside the country. It also bans *Michelia* sp. (Champ), *Acacia catechu* (Khayer) and *Shorea robusta* (Sal) for transportation, export and felling. Recently HMG, Nepal lifted the ban on the collection of *Cordyceps sinensis* with effective from February 12, 2001. However, another decision made at the same time added *Neopicrorhiza scrophularifolia* in the list of banned herb.

Plants protected in Nepal are observed effective in nine national parks, four wildlife reserves and two world heritage sites. Plants are also protected in one hunting reserve and one Ramsar wetland.

In spite of all these measures taken at national and international levels genetic erosion of various plant species continues in Nepal. Identification, domestication and multiplication of medicinal and aromatic plants of commercial values, however, believed to add effectiveness in conservation, protection and expansion of national natural heritage leading to the socio-economic development of local communities.

Plants included in this work are found wild in Nepal. A few of them are naturalized in remote past. Those plants that are largely used in Nepal by local communities for medicinal purposes and also for Ayurvedic preparations but not reported in wild are omitted. Only accepted names are given as the title for each taxon in the main text. Press *et al.* (2000) has been followed for names, Synonym(s), and the general distribution of the taxon within and outside Nepal In addition to the literature available, personal communications were also used for the information on common name(s) and medicinal and aromatic uses. In some cases, local names as Dolpali Nepali and those used by Amchis (Ghimire *et al.* 2000) have been used. Amchis are known as the local healers mostly influenced by Tibetan traditional medicine.

Abies spectabilis (D.Don) Mirb.

Synonym(s): *Abies densa* Griff. ex R. Parker; *A. webbiana* Lindl.
Common name(s): Gobre sallo/Talispatra (Nep.); Himalayan silver fir (Eng.)
Family: Pinaceae

Description: An evergreen much branched tree of up to 50 m or more in height but often a gnarled and much smaller tree with upper branches horizontal, and branchlets horizontal and flattened, and with dark gray, deeply longitudinally grooved bark. Distinguished by the young branchlets which are hairy in the grooves but soon become hairless, and by the leaves which are distinctly parted on the upper sides of the branchlets and point sideways, and with the apex of leaf rounded and notched. Leaves flattened, up to 4 cm long, pale beneath with incurved margins. Cones dark purple, erect, cylindric, 10-20 cm long and 4-7.5 cm broad; male catkins cylindric *c.* 5 cm.

Distribution: WCE; alt. 2400-4400 m.
Pakistan, Himalayas (Kashmir to Nepal).
(A. spectabilis var. *langtanensis* Silba) C; 3048 m. Nepal.

Part(s) used: Leaves.

Important biochemical constituent(s): White resin.

Uses: Leaves used as carminative, expectorant, stomachic, tonic, astringent, in asthma and bronchitis.

Abrus precatorius L.

Common name(s): Rati gedi, Lal gedi (Nep.); Gunja (Sans.); Crab's eye, Jamaica wild liquorice (Eng.)
Family: Leguminosae
Chromosome number(s): 2n=22

Description: A climbing shrub with numerous slender branches. Leaves compound 4.5-7.5 cm; leaflets 10-20 pairs. Flowers pink or blue 2.5-7 cm long in cluster. Pod 2.5-5 cm long with 3-6 seeds red and black hard and shining. Each dry seed weighs approximately 125 mg.

Distribution: WE; alt. 300-1100 m.

Dr. Kamal K. Joshi and Prof. Sanu Devi Joshi

Tropics and subtropics of Africa, Asia, south to Australia, Pacific Islands.

Part(s) used: Fruits and seeds.

Important biochemical constituent(s): The plant contains glycyrrhizin and the seeds taxalbumin "abrin".

Uses: Fruits are bitter, acrid, aphrodisiac, useful in eye disease, care leucoderma. Seeds are purgative, but in large doses are an acrid poison, also use to prevent conception (Kirtikar and Basu 1987, Malla *et al.* 1997).

Abutilon indicum (L.) Sweet

Synonym(s): *Abutilon indicum* var. *populifolium* (Lam.) Wight & Arn. ex Mast.
Common name(s): Kagiyo (Nep.); Atibala (Sans.)
Family: Malvaceae
Chromosome number(s): 2n=36, 42

Description: A softly tomentose perennial shrub to about 3 m tall with round stem tinged with purple. Leaves ovate to orbicular-cordate. Flowers golden yellow, solitary on jointed peduncles often forming a panicle-like terminal inflorescence. Fruits hispid, scarcely longer than calyx; awns erect; seeds 3-5, kidney-shaped, dark brown or black.

Distribution: CE; alt. 200-1100 m.
Himalayas, India, Sri Lanka, Myanmar, Thailand, S. China, Taiwan, S. Japan, Malaysia, Australia, Loyalty Islands.

Part(s) used: Whole plant.

Important biochemical constituent(s): The plant is reported to contain n-alkane mixture (C_{22}-C34), β-sitosterol, ρ-coumaric acid, ρ-hydroxybenzoic acid, ρ-β-D-glycosyloxybenzoic acid, leucine, histidine, threonine, serine etc (Kirtikar and Basu in Anonymous 1985).

Uses: The plant is used as diuretic, demulcent, laxative and antipyretic. An infusion is prepared from roots, which is given in fevers as a cooling remedy. It is also useful in the treatment of leprosy. The bark is astringent and used as diuretic.

Leaf paste is used to aid in healing cut and wounds. A decoction of the leaves is used in gonorrhea and chronic bronchitis. The leaves are cooked and eaten in cases of bleeding piles.

The seeds are laxative and demulcent, and are used in the treatment of cough. The seeds are aphrodisiac.

Acacia catechu (L.f.) Willd.

Synonym(s): *Mimosa catechu L.f.; M. catechoides* Roxb.
Common name(s): Khayer (Nep.); Khadira (Sans.); Cutch tree, white catechu (Eng.)
Family: Leguminosae
Chromosome number(s): 2n=26

Description: A medium-sized tree of up to 9-12 m in height with spines. Leaves bipinnate, 10-15 cm long with 30-50 pairs of leaflets. Flowers pale yellow or white in peduncled axillary spikes. Pods 5-7.5 cm long with 3-10 seeds.

Distribution: WCE; alt. 200-1400 m.
Tropical Himalayas, India Myanmar, Thailand, S. China.

Conservation status: Commercially threatened (IUCN Category). **Part(s) used:** Bark and heartwood.

Important biochemical constituent(s): Catechin, catechutannic acid, glucose and sucrose.

Uses: Used in chronic diarrhea, dysentery, ulceration of mouth, particularly the gum, obstinate skin diseases. The resinous extract in powder is given for drying wounds. It is used to kill worms in cattle. Kattha (the extract) is used in pan. Astringent, cooling, digestive beneficial in cough and diarrhea, applied externally to ulcers, boils and eruption of the skin (Rajbhandari *et al.* 1995).

Acacia nilotica (L.) Willd. ex Delile

Synonym(s): *Acacia arabica* (Lam.) Willd.; *Mimosa arabica* Lam.; Mimosa nilotica L.
Common name(s): Babbur (Nep.); Babul (Eng.)
Family: Leguminosae
Chromosome number(s): 2n=26, 52, 104

Description: An evergreen tree of 6-9 m in height with spines. Spines straight, up to 4 cm long. Leaves with glands; leaflets 10-20 pairs. Inflorescence yellow, aromatic, around 1 cm in diameter. Pod 7.5-15 cm long and 1.5 cm broad, flattened and having 8-10 seeds.

Distribution: C; alt. 150 m. Nepal India, Sri Lanka.

Part(s) used: Bark and resinous extract.

Important biochemical constituent(s): Gum contains arabic acid with calcium, magnesium and potassium and sugars (Anonymous 1994).

Uses: Used in disease of mouth, diarrhea, gonorrhoea and weakness.

Acacia rugata (Lam.) Voigt.

Synonym(s): *Acacia concinna* (Willd.) DC.; *Mimosa concinna* Willd.; *Mimosa rugata* Lam.
Common name(s): Sikakai, Rasula (Nep.)
Family: Leguminosae

Description: A climbing shrub with prickles. Leaves bipinnate, 5 cm long, rachis with a large gland at or below the middle of the petiole and one between the uppermost or two uppermost pairs of pinnae, stipules ovate, cordate, pinnae 4 pairs, 2.5-6.5 cm long. Leaflets sub-sessile, sensitive, 12-25 pairs, 0.6-1.3 cm long and 0.13 cm wide, linear, acute or mucronate, unequal sided. Flowers 0.3-0.4 cm long in globose heads on 2.5-4 cm long peduncles which are fascicled at the nodes or forming panicles at the end of the branches. Pods linear-oblong, 7.5-12.5 cm long, 2-3 cm broad with 6-10 seeds (Malla *et al.* 1997).

Distribution: WC; alt. 400-800 m.
Nepal, India, SE Asia, S. China, Malaysia.

Part(s) used: Leaves and pods.

Important biochemical constituent(s): Large amount of saponin and tannin have been isolated from the pods.

Uses: Pods are used as expectorant, purgative, anthelmintic, antidiarrheal and emetic. The leaves are used as cathartic and in biliousness. Pods and leaves are used for hair growth and malarial fever (Malla *et al.* 1997).

Achyranthes aspera L.

Common name(s): Datiwan, Apamarga (Nep.); Apamarga (Npbh.); Apamarga (Sans.); Prickly or rough chaff flower (Eng.)
Family: Amaranthaceae
Chromosome number(s): 2n=14, 21, 36, 42, 84.

Description: An erect herb or undershrub of up to 1 m in height. Stem stiff with pubescent, terete and striate branches. Leaves few, thick, elliptic or obovate, pubescent, usually rounded at the apex. Flowers green or tinged with purple, numerous, on a long and woody spike.

Distribution: WCE; alt. 100-2900 m. Pantropical

Part(s) used: Whole plant.

Uses: Pungent, purgative, diuretic, astringent, used in dropsy, and piles. Leaf juice is used in stomach ache, piles, skin eruptions. Roots for pyorrhea and cough (Ahuja 1965).

The plant is reported to possess antidiabetic and antirheumatic properties and used beneficially in abdominal tumors. Roots are used as stomachic and digestant and are said to be useful in the treatment of pneumonia as well as toothache. Extract of the root is used to treat menstrual disorders and dysentery. Root paste is used to be an antifertility drug. Leaf is used in boil, abscess, stomachache, bowel complaints, piles and skin eruptions. Paste of the leaf is used to treat bites of poisonous insects, wasps and bees. Seeds are used to treat snakebites, hydrophobia and itching. Seed powder is used in the treatment of bleeding piles. Seeds are emetic and also used as a brain tonic Decoction of the whole plant is diuretic and used in renal dropsy and generalized anasarca. It is also given in painful delivery. Juice of the plant is used to stop bleeding of wounds (Bhattacharjee 1998).

Achyranthes bidentata Blume

Common name(s): Datiwan, Apamarga (Nep.); Apamarga (Npbh.); Prickly chaff flower, two-toothed chaff flower (Eng.)
Family: Amaranthaceae
Chromosome number(s): 2n=24, 36, 42, 84.

Description: Pubescent herb of around 30 cm in height. Leaves opposite, simple, petiolate, exstipulate, elliptic to ovate, acute, 2-8 cm long. Flowers in slender spike, 4-6 cm long, sepals 4, lanceolate and shining. Bracts usually reduced to spine, 2-auricled at the base. Stamens 5, staminodes toothed. Ovary sub-compressed, oblong. Style filiform. Stigma capitellate (Malla *et al.* 1986).

Distribution: CE; alt. 1200-2100 m.
Tropical Africa, Himalayas, India, East to China, Malaysia.

Part(s) used: Root, stem, leaves and seeds.

Uses: Diuretic and astringent. White variety bitter, pungent, heating, laxative, stomachic, itching and in pain in the abdomen, ascites, dyspepsia, dysentery. Seeds are useful in piles. Red variety pungent, cooling, emetic, constipating, alexipharmic, dried plant is given in colic (Parajuli *et al.* 1998). Roots are used in sore throat, hypertension, amenorrhea, retention of placenta, carbuncles, traumatic injury, asthenia of liver and kidney, tiredness in lower part of the body and the legs and in rheumatic pain (WHO 1989).

Aconitum bisma (Buch.-Ham) Rapaics

Synonym(s): *Aconitum ferox* subsp. *palmatum* (D.Don) Brühl, *Aconitum palmatum* D.Don and *Caltha bisma* Buch.-Ham.
Common name(s): Bikhma (Nep.); Aconite (Eng.)
Family: Ranunculaceae
Chromosome number(s): 2n=30

Description: A glabrous herb with tuberous roots. Leaves glabrous or pubescent above, blade cordate to reniform with wide sinus, 3- or 5-partite above, divisions obovate-cuneate to broadly lanceolate-cuneate or lateral trapezoid, 3-lobed middle segment and 2-lobed outer segment, intermediate lobe elongated, lobes inciso-dentate or apiculately creneate. Inflorescence loose, leafy panicle or raceme, 10-20 cm long, glabrous or pubescent above; bracts ovate, deltoid, dentate; bracteoles smaller, sparingly dentate or entire. Sepals bluish white or variegated, posterior sepal helmet-shaped, shortly or obscurely beaked. Petals glabrous; claw erect or the tip more or less leaning forward, hood sub-cylindric, oblique to horizontal, gibbous with honey gland. Seeds obovoid, obscurely winged.

Distribution: C.
Himalayas (Nepal, Arunachal Pradesh), China (Xixang)

Part(s) used: Roots.

Uses: The root is bitter, nonpoisonous, and used as a tonic. It is also used in combination with long pepper *(Piper longum* L.) for pain in the bowels, for diarrhea, and vomiting. Externally, it is used as an application for rheumatism, cuts and wounds.

Aconitum ferox Wall. ex Ser.

Common name(s): Bikh (Nep.); Aconite (Eng.) **Family:** Ranunculaceae
Chromosome number(s): 2n=20, 34

Description: A perennial herb 1-2 m tall with tuberous roots. Leaves with round or oval blade 8-15 cm across having 5-wedge-shaped lobes which are further deeply lobed with ultimate segments at least 3 mm wide. Flowers dull blue in a terminal spike-like cluster 15-30 cm long simple, sparingly branched below, hood 2-3 cm broader than long, with a short acute beak. Lower bracts of inflorescence pinnately lobed, upper entire; bracteoles linear. Follicles 5, usually finely hairy.

Distribution: CE. alt. 2100-3800 m. Himalayas (Nepal, Bhutan).

Conservation status: Commercially threatened (IUCN Category).
Part(s) used: Tuberous root.

Uses: The plant is poisonous and used in many preparations of traditional medicines. It can also be used as analgesic, antipyretic, diaphoretic, natural rodenticide and insecticide.

Dr. Kamal K. Joshi and Prof. Sanu Devi Joshi

Aconitum gammiei Stapf.

Common name(s): Nirmashi (Nep.)
Family: Ranunculaceae

Description: A weak flexuous herb to 1 m tall. Leaves few, 4-5 cm across, deeply lobed; lobules narrow, acute. Flowers pale blue, white or greenish yellow in a loose few-flowered inflorescence.

Distribution: CE. alt. 3300-4300 m. Himalayas (Nepal, Sikkim), China (Xizang)

Conservation status: Rare (IUCN Category).
Part(s) used: Tubers.

Uses: Non-poisonous and used in variety of traditional medicines (Shrestha and Joshi 1996).

Aconitum heterophyllum Wall.

Common name(s): Atis (Nep.); Ativisha (Sans.); Aconite (Eng.) **Family:** Ranunculaceae
Chromosome number(s): $2n = 16$

Description: An erect leafy herb found in wet and open place and in riverbed forest, up to 1 m tall. Leaves broadly ovate or disk-heart-shaped; lower leaves long pitioled, more or less divided into 5 segments and thoothed; upper leaves non-divided and amplexicauled. Flowers bright blue to greenish blue with purple veins, in loose spike-like cluster (Shrestha and Joshi 1996).

Distribution: C; alt. 3200-3700 m. Himalayas (Kashmir, Nepal).

Conservation status: Rare (IUCN Category). **Part(s) used:** Root.

Important biochemical constituent(s): Root contains bitter alkaloid atisine, acotinic acid and pectinous substances (Anonymous 1994).

Uses: Non-poisonous. Root exhibits varieties based on the color viz., white, yellow, red and black. Plant with white roots is considered the best for medicinal use. Roots are bitter tonic, stomachic, and digestive. They alleviate dysentery and bilious complaints and are good in periodic and intermittent fevers, dyspepsia and cough (Shrestha and Joshi 1996; Malla *et al.* 1998).

Aconitum laciniatum (Brühi) Stapf

Synonym(s): *Aconitum ferox* var. *laciniata* Brühi **Common name(s):** Kalo bikh (Nep.); Nepal aconite (Eng.) **Family:** Ranunculaceae
Chromosome number(s): 2n=40

Description: An erect herb with stiff or flexous stem to 50 cm tall. Leaves with slender petiole and dissected lamina, sinuses usually wide and shallow with 5-pediti-partite, incesodented or laciniate lobes. Leaf blades often finely pubescent, inflorescence loosely paniculate, few to many-flowered. Flowers blue or red purple, finely pubescent on sepals. Nectaries hispidulous. The tuberous rhizomes are highly poisonous (Shrestha and Joshi 1996).

Distribution: CE; alt. 3800-4800 m. Himalayas (Nepal, Bhutan), China (SE Xizang)

Conservation status: Commercially threatened (IUCN Category).
Part(s) used: Tuberous rhizomes.
Important biochemical constituent(s): Aconite.

Uses: Tuberous rhizomes are highly poisonous. Used in a variety of traditional medicines.

Aconitum spicatum (Brühl) Stapf

Synonym(s): *Aconitum ferox* var. *spicata* Brühl
Common name(s): Bikh (Nep.); Aconite (Eng.) **Family:** Ranunculaceae
Chromosome number(s): 2n=32

Description: A perennial herb to 2 m high. Leaves often softly hairy, blades 6-12 cm across deeply lobed, lobes ovate and toothed. Flowers in dense terminal spikes. Occasionally with lateral spikes, purple to greenish white, hooded.

Dr. Kamal K. Joshi and Prof. Sanu Devi Joshi

Distribution: WCE; alt. 1800-4200 m. Himalayas (Nepal to Bhutan), China (Xizang).

Conservation status: Commercially threatened (IUCN Category).
Part(s) used: Tubers

Important biochemical constituent(s): Bikhaconitine

Uses: Tubers deadly poisonous, antipyretic and analgesic. Used in variety of traditional medicines.

Aconogonum rumicifolium (Royle ex Bab.) Hara

Common name(s): Chaunle (Dolp. Nep.) **Family:** Polygonaceae
Chromosome number(s): 2n=20

Description: A perennial herb with large dock-like leaves and usually dense axillary and terminal clusters of tiny green flowers. Flowers 4-6 mm across with rounded spreading perianth-segments. Leaves fleshy, broadly ovate to ovate heart-shaped, 8-13 cm, blunt or almost acute, leaf stalk short, stout, to 2.5 cm; stipules large, lax, hairless; stem stout, unbranched, pale, 15-120 cm (Polunin and Stainton 1986).

Distribution: WC; alt. 3300-4400 m.
Afghanistan, Himalayas (Kashmir to Nepal), China (Xizang).

Part(s) used: Root, stem and leaf.

Uses: Decoctions of root, stem and leaf are given in abdominal pain and dysentery.

Acorus calamus L.

Common name(s): Bojho (Nep.); Säpi (New.); Vacha (Sans.); Sweet Flag (Eng.)
Family: Araceae
Chromosome number(s): 2n=18, 24, 36, 48

Description: An erect herb with aromatic rootstock. Leaves 15-35 cm long and 0.5-1 cm broad, ensiform, with distinct midrib and wavy margins. Peduncle 0.2-0.5 cm broad. Spathe 3.5-4 cm long and 0.3-0.5 cm in diameter, cylindric, slightly curved. Flowers bisexual, each with a perianth of 6 orbicular concave segments, stamens 6, ovary conical, 2-3-celled. Fruits oblong berries 4 cm long.

Distribution: WCE; alt. 1700-2300 m. Temperate N. Hemisphere

Part(s) used: Rootstocks and rhizomes.

Important biochemical constituent(s): Volatile oil "Asaryl - aldehyde" and bitter glucoside "Acorin".

Uses: Root stock is emetic, stomachic, used in dyspepsia, colic, remittent fever, bronchitis, dysentery and chronic diarrhea. It is also used as nerve tonic. Rhizomes are used as carminative, stimulant and tonic. An essential oil is extracted from the rhizomes to use in perfumes.

Aegle marmelos (L.) Correa

Common name(s): Bel (Nep.); Byaha (Npbh.); Bilva (Sans.); Bengal quince (Eng.)
Family: Rutaceae
Chromosome number(s): 2n=18

Description: A deciduous spiny glabrous tree, about 10 m high. Leaves odd pinnate, trifoliate, leaflets 3, rarely 5, ovate-lanceolate, lateral leaflets nearly sessile and the terminal one petiolate, 4-6 cm long, crenate, obtuse, cuneate. Flowers in short panicles, 2 cm in diameter, white, sweet-scented. Sepals 5-lobed, deciduous. Petals 5, thick, oblong, much longer than the sepals, gland dotted. Stamens numerous; filaments short, subulate; anthers narrow, elongated, basifixed, erect. Fruit gray or yellow, berry, globose, 3-9 cm in diameter, rind woody. Seeds numerous, flat, oblong, woolly, lying embedded in aromatic pulp

Distribution: WCE; alt. 600-1100 m.
Himalayas (Kashmir to Nepal), India, Myanmar, Indo-China, Malaysia. Widely cultivated.

Part(s) used: Ripe fruit

Important biochemical constituent(s): Ripe fruit contains tannin 9% in the pulp and 20% in the rind.

Uses: Ripe fruit is laxative, used against constipation and dyspepsia. Unripe fruit is astringent, digestive and used against diarrhea. Root and bark are used against fever and also for the preparation of 'Dasmula ', an ayurvedic preparation (Ahuja 1965; Rajbhandari *et al.* 1995 and Malla *et al.* 1997). Roots, leaves and fruits have antibiotic properties. Fruit is used as drink and medicine. Unripe fruit is powdered and taken with water in dysentery. Fruit is valuable for its mucilage and pectin (Bhattacharjee 1998).

Agrimonia pilosa var. nepalensis (D.Don) Nakai

Synonym(s): *Agrimonia eupatorium* auct; *A. lanata* Wall. ex Wallr.; *A. nepalensis* D. Don
Common name(s): Hairy agrimony (Eng.)
Family: Rosaceae
Chromosome number(s): 2n=56

Description: A perennial herb to 100 cm tall with brownish, woody horizontal rootstock having fibrous roots. Leaves interruptedly pinnate; leaflets 7-21, sessile, hairy on both surfaces and punctate beneath; larger ones 1.5-3 cm, elliptic-ovate or obovate, rarely orbicular, deeply dentate, irregular and smaller ones often orbicular and minute, petiole slender, stipules large, leafy, incised-dentate, partially adnate to the petiole. Inflorescence an elongated terminal raceme. Flowers small, yellow. Fruit an achene enclosed within the hard spinose calyx.

Distribution: WCE; alt. 1000-3000 m.
Himalayas (Kashmir to Bhutan), NE India, Myanmar, China.

Part(s) used: Whole plant.

Uses: Decoction of whole plant (6-12 g) is given to cure gastrorrhagia, haematemesis, epistaxis, haematuria, melaena, dysentery, vaginal trichomoniasis (external use only). Powdered winter buds (30 g for adults) are given in tapeworm infestation (Anonymous 1989).

Albizia julibrissin Durazz. Var. julibrissin

Common name(s): Silk tree (Eng.); Shirish (Nep.)
Family: Leguminosae
Chromosome number(s): 2n=26, 52
Description: A deciduous tree to 16 m high with dark gray bark. Young shoots and inflorescence clothed with yellowish brown pubescence. Leaves bipinnate, stipules 7.5 mm, long, linear, caducous. Pinnae 4-16 pairs, to 15 cm long, leaflets 10-30 pairs, 1.3-1.8 cm long and 0.4-0.6 cm broad. Flowers pink in peduncled head, solitary or in fascicles of 2-3 arranged in a short terminal raceme. Pod 9.15 cm long and 1.5-2.5 cm broad, thin, pubescent, glabrous when mature, pale brown or yellowish with 8-12 seeds.

Distribution: E; 1300-1400 m.
Tropical to warm temperate Africa, Asia; often cultivated.

Part(s) used: Bark, flowers or flower buds.

Uses: Bark is given to treat anxiety and insomnia, pulmonary abscess, cough and trauma. Flowers or flower buds are effective in insomnia, amnesia and feeling of constriction in the chest (Anonymous 1989).

Albizia lebbeck (L.) Benth

Synonym(s): *Mimosa lebbeck* L.; *M. sirissa* Roxb.
Common name(s): Shirish (Nep.); Sirisha (Sans.)
Family: Leguminosae
Chromosome number(s): 2n=26

Description: A deciduous tree to 21 m. tall with pale bark and glabrous young shoots. Leaves bipinnate, with a large gland on the petiole above the base and one below the uppermost pair of pinnae; pinnae 2-3 (rarely 4) pairs, 10-12.5 cm long. Leaflets 5-9 pairs, 2.5-4.5 long and 1.6-2 cm broad, with gland between the bases, lateral leaflets elliptic-oblong and terminal 2 obovate-oblong. Flowers white, fragrant, in globose umbellate heads 2-3.8 cm diam.; peduncles 3.8-7.5 cm long, solitary or 2-4 together from the axils of the upper leaves. Pods 10-30 cm long 2-4.5 cm broad, linear-oblong. Seeds 4-12, ellipsoid-oblong, compressed, light brown.

Distribution: WCE; alt. 250-800 m.
A native of tropical Himalayas, India, Sri Lanka, SE Asia, S. China; cultivated widely in tropics and subtropics.

Part(s) used: Bark, seeds and flowers.

Important biochemical constituent(s): Tannins and pseudotannins, friedelin and γ-sitosterol have been isolated from the bark. Seeds contain saponins based on echinocystic acids (Anonymous 1994).

Uses: In ayurvedic medicine, decoction of bark and powdered seeds (Shirisharisht or Shirishadi) are given in the morning and the evening before meal to treat toxicity and as blood purifier. As a remedy for breathing problems, it is given mixed with rhizomes of *Curcuma longa* or fruits of *Piper longum* and honey 2-3 times a day. Paste of bark or seeds (Dashang) is applied locally in the treatment of skin diseases (Adhikari 1998).

Allium carolinianum DC.

Synonym(s): *Allium aitchisonii* Boiss.; *A. blandum* Wall.; *A. obtusifolium* Klotzsch.; *A. polyphyllum* Kar. & Kir.; *A. thomsonii* Baker
Common name(s): Jangali lasoon, Kage lasoon (Nep.)
Family: Amaryllidaceae
Chromosome number(s): 2n=16, 32

Description: A stout herb to 30 cm tall with a very dense globular umbel of pink flowers. Leaves glaucous, usually shorter, broad, flat and curved. Umbels 2-3.5 cm across, flowers cylindrical; petals elliptic, pointed, to about 6 mm, much shorter than the stamens. Bulbs relatively large, oblong-cylindric, covered with conspicuous leathery scales.

Distribution: WC; alt. 4800-5100 m.

C. Asia, Afghanistan, Himalayas (Kashmir to Nepal).

Part(s) used: Roots and leaves.

Uses: Tonic. Aids in digestion. Also used to relieve from toothache, earache and headache.

Allium fasciculatum Rendle

Synonym(s): *Allium gageanum* W. W. Sm.
Common name(s): Faran (Dolp.) **Family:** Amaryllidaceae
Chromosome number(s): 2n=20

Description: A bulbless herb with cluster of many stout tuberous roots. Stem as long as or shorter than the leaves, flowering stem usually 12-20 cm long, sometimes more. Leaves numerous, 3-4 mm wide. Inflorescence a lax rounded umbel. Flowers sweet-scented, white or pale green, *c* 5 mm; petals lanceolate, pointed, shorter than the stamens.

Distribution: WC; alt. 2800-4500 m. Himalayas (Nepal to Bhutan), China (Xizang).

Part(s) used: Whole plant.

Uses: Stimulant. Flavoring material to lentil soups and many other recipes. Used as aids in healing wounds and curing swelling, toothache and sore throat.

Allium wallichii Kunth

Synonym(s): *Allium violaceum* Wall. ex Regel
Common name(s): Ban lasoon (Nep.)
Family: Amaryllidaceae
Chromosome number(s): 2n=16, 32

Description: A glabrous bulbless herb numerous fibrous roots and 3-angled stem, to 90 cm tall. Leaves many, linear, or spear-shaped, flat and keeled, to 2 cm broad, often as long as the stem. Flowers purple, numerous, long-stalked, in a lax rounded umbel, 5-7 cm across. Petals broadly linear, longer than the purple stamens and ovary.

Distribution: WCE; alt. 2400-4650 m. Himalayas (Nepal to Bhutan), W. China.

Part(s) used: Fibrous roots and leaves.

Uses: Stimulant and flavoring material to various recipes.

Alnus nepalensis D. Don

Common name(s): Utis (Nep.); Bosin (Npbh.); Alder (Eng.)
Family: Betulaceae
Chromosome number(s): 2n=56

Description: A deciduous perennial tree to 33 m high with root bearing nodules of *Frankia alni* enabling the tree to fix atmospheric nitrogen in the soil. Bark glabrous gray. Leaves simple, lanceolate, elliptical, serrate, lateral veins on the upper part curved, Flowers in catkin; male catkins in terminal drooping panicles, female catkins cone-like one or several at the axils. Fruits numerous, ellipsoid or sub-cylindric; nuts scarcely with wings.

Distribution: WCE; alt. 500-2600 m.
Himalayas (Uttar Pradesh to Bhutan), NE India, Myanmar, Indo-China, W. China.

Part(s) used: Roots and leaves.

Uses: A decoction of the root is taken orally for the treatment of diarrhea and the leaf paste is applied in cuts and wound.

Aloe vera (L.) Burm. f.

Synonym(s): *Aloe barbadensis* Mill.; *A. indica* Royle; *A. vulgaris* Lam. **Common name(s):** Ghiukumari, Musabar (Nep.); Kunuh (Npbh.); Kumari (Sans.); Indian Aloe (Eng.)
Family: Liliaceae
Chromosome number(s): 2n=14

Description: A perennial herb with short stem and rosulate leaves to 50 cm long and 8 cm wide, rounded on the reverse, gray green. Inflorescence receme, flowers cylindric, yellow.

Distribution: W; alt. 1200-1400 m.
Mediterranean & Canary Islands, Naturalized in Florida, West Indies, Central America & Asia.

Part(s) used: Leaves

Important biochemical constituent(s): Anthraquinone glycosides collectively termed as "aloin". In *A. barbadensis,* the aloin content is 30 per cent.

Uses: Stomachic, cooling, alterative, purgative and emmenagogue, used against piles and rectal fisures. Useful in eye disease, tumors, enlargement of the spleen, liver complaints, vomiting. The mucilage is cooling and used to poultice inflammations and extensively used in cosmetic preparations (Malla *et al.* 1997).

Alstonia scholaris (L.) R. Br.

Common name(s): Chhatiwan (Nep.); Chhatiwansin (Npbh.); Saptaparna (Sans.); Dita bark (Eng.)
Family: Apocynaceae

Chromosome number(s): 2n=40, 44

Description: A tall evergreen tree of 25 m in height with a small trunk and spreading whorled branches having bitter milky juice. Bark gray, uneven and rough, with many lenticels on the younger branches. Leaves in whorls of 5-7, to 20 cm long, lanceolate or obovate oblong, midrib very prominent beneath, leathery. Flowers small, greenish white short-stalked, in small clusters combined into compact long peduncle, rounded, paniculate cymes. Fruits cylindric, pendulous, narrow, slender, to 60 cm long, hanging in pairs forming dense clusters. Seeds dorsally compressed, numerous, pale orange.

Distribution: CE; alt. 100-300 m. Tropical Asia and Africa.

Conservation status: Rare (IUCN Category).
Part(s) used: Bark.

Important biochemical constituent(s): Diamine (an alkaloid).

Uses: The bark has bitter taste, not acrid and devoid of any odor. An alkaloid has been detected from the bark which is known as diamine (Bhattacharjee 1998).

Amaranthus spinosus L.

Common name(s): Ban lunde, kande lunde (Nep.); Kanbakan (Npbh.); Alpamarisha, Tandula (Sans.); Prickly amaranth, Spiny amaranth (Eng.)
Family: Amaranthaceae
Chromosome number(s): 2n=32, 34
Description: A prickly annual herb to about 75 cm tall, stem erect angular and hollow. Leaves simple, long petioled, narrowly or sometimes broadly ovate, 3-6 cm long and 1.5 to 4 cm broad with axillary spines. Flowers sessile in axillary and terminal and cylindric spikes, greenish, each with a perianth of 5 narrowly oblong, membranous, cuspidate segments. Fruits ovoid, 1.5 mm long, imperfectly circumscissile.
Distribution: WCE; alt. 150-1200 m. Cosmopolitan, warm temperate and tropical weed.

Part(s) used: Root.

Uses: Diuretic, also used in colic pain and leucorrhoea.

Amomum subulatum Roxb.

Common name(s): Alainchi (Nep.); Ela (Npbh.); Aindri, Sthulaela (Sans.); Greater cardamom (Eng.)
Family: Zingiberaceae
Chromosome number(s): 2n=48 (26, 32, 34, 42, 44)

Description: An aromatic crop. Dried ripe fruits give a strongly flavored spice. Fruit a capsule, ovoid or oblong, 1 - 2 cm long, 0.5 - 1.8 cm broad, externally longitudinally straited and dark brown in color. Seeds dark brown, more or less round and 3-4 mm in diameter (Rajbhandari *et al.* 1995).

Distribution: CE; alt. 1000-2000 m.
E. Himalayas (Nepal to Sikkim), N. India. Cultivated on shady slopes.

Part(s) used: Seeds.

Important biochemical constituent(s): Terpenes in volatile oil from fruits. Oil content is 2.043% (Rajbhandari *et al.* 1995).

Uses: Seeds - stomachic, useful in neuralgia. Oil from seeds - aromatic, stimulant, stomachic, appetizer and applied to eyelids to allay inflammation.

Ampelopsis japonica var. mollis (Wall ex M. A. Lawson)

Common name(s): Japanese ampelopsis (Eng.) **Family:** Vitaceae
Chromosome number(s): 2n=40

Description: A perennial climbing vine about 1 m long with fleshy tuberous roots. Tubers in clusters, robust, long-fusiform or ovate, dark brown externally and white starchy inside. Stem ligneous at the base, much branched, young branches glabrous, finely striate, slightly purplish. Tendrils opposite to the leaves. Leaves compound, palmate, 6-10 cm long and 7-12 cm wide, leaflets 3-5, pinnatilobate, rachis broadly winged. Flowers small, yellowish green. Fruit berry, globose, blue or bluish-purple at maturity.

Distribution: E; alt. 1600-1700 m.
Himalayas (Nepal to Bhutan), NE India, Bangladesh, ?Malaya.

Dr. Kamal K. Joshi and Prof. Sanu Devi Joshi

Part(s) used: Tuberous roots.

Uses: Tuberous roots are sweet and are used in the treatment of skin inflammation, pyogenic infection, ulcerous disease of skin. In skin burns the tuber is used externally (Anonymous 1989).

Anagallis arvensis L.

Family: Primulaceae
Chromosome number(s): 2n=40

Description: An annual herb attaining to 38 cm tall, erect or procumbent. Leaves to 2.5 cm long, sessile, opposite, ovate, glabrous, entire, gland dotted. Flowers blue, axillary, solitary, peduncles to 3.8 cm long, slender, erect in flower and decurved in fruit. Fruit a capsule, small of the size of a small pea. Seeds minute, trigonous.

Distribution: WC; alt. 600-2700 m.
Europe, W. Asia, most N. Temperate regions, Himalayas (Kashmir to Nepal), India, Sri Lanka.

Part(s) used: Whole plant.

Uses: It is used as expectorant in cases of lung abscess.

Androsace rotundifolia Hardw.

Family: Primulaceae
Chromosome number(s): 2n=20

Description: A small perennial hairy tufted herb having long-stalked rounded leaves. Flowers pink and white in irregular umbel.

Distribution: WC; alt. 910-3700 m.
Afganistan, Pakistan (Chitral), Himalayas (Punjab to Bhutan).

Part(s) used: Leaves.

Uses: Leaves are eaten with molasses or sugar in irregular menstrual flow in age group of 25-30 years (Bhattacharjee 1998).

Apium graveolens L.

Common name(s): Ajmoda (Nep.); Ajmoda (Sans.), Celery (Eng.) **Family:** Umbelliferae
Chromosome number(s): 2n=22

Description: Annual herb with adventitious roots and erect stem. Leaves compound, pinnate with long stalks. Flowers greenish white in compound umbels. Fruits with very small dark brown cremocarp seeds with pungent taste and agreeable odor.

Distribution: C; alt. 2500-2800 m.
Nepal. Widely distributed throughout the world, often as a result of escape from cultivation.

Part(s) used: Leaf, stalk, seed and its oil.

Important biochemical constituent(s): The oil of herb constitutes apiol, sedanolide and 3-butylphalide. The oil of seed contains d-limonene, d-selenene, selanoic acid anhydride and sedamolide. The leaves and the stalks contain vitamin A, C and iron. The herb is also reported to contain glucoside appin (Bhattacharjee 1998).

Uses: The essential oil of seeds is used in pharmaceutical industries and for flavoring food. Root is used as an aperitive and diuretic and administered for the treatment of jaundice, nephritic colic and obstructions in the urinary passages (Bhattacharjee 1998).

Arachis hypogaea L.

Common name(s): Badam (Nep.); Baran (Npbh.); Peanut (Eng.)
Family: Leguminosae
Chromosome number(s): 2n=40

Description: A bushy or creeping annual herb with stem cylindrical, becoming more or less angular with age, hairy, hairs 3 types, long, short and glandular. Leaves pinnate with two pairs of leaflets borne on a slender, grooved petiole. Flowers yellow, papilionate, sessile in the axils of leaves on short branches. Corolla lemon yellow. Stamens 10, monadelphous, with dimorphic anthers. Fruit a pod, ripening underground.

Distribution: C; alt. 800-1000 m. Cultivated widely throughout the world.

Part(s) used: Oil extracted from the seeds.

Uses: British Pharmacopoeia recognizes groundnut oil as a substitute for olive oil in the preparation of official liniments, plasters, ointments and soaps. The oil has been declared suitable for making medicine for the injection of bismuth salicylate and mercury. The oil can be used for preparing concentrated solutions of vitamins A and D (Samba Murty and Subrahmanyam 1989).

Arctium lappa L.

Common name(s): Kurro (Dolp.); Great burdock, Clotbur, Cocklebur (Eng.)
Family: Compositae
Chromosome number(s): 2n=32, 36

Description: A biennial or perennial herb to 2 m. high. Stem erect, sturdy, striate, branching, purplish, puberulent. Leaves towards base petiolate, often cordiform, 40-50 cm long and 30-40 cm broad, obtuse, base cordate, margin undulate, upper surface glabrous, lower surface covered with cottony white hairs and leaves towards apex oval. Flowers arranged in corymbs at the tips of the branches, purple; corolla tubular, 5-lobed. Fruit an achene, long, elliptic or obovate, slightly traingular, 5-6 mm long and 2.5 mm broad, surface grayish brown with black spots.

Distribution: WC; alt. 2100-3700 m. Eurasia

Part(s) used: Fruits

Uses: Fruits are pungent and used in common cold, cough, headache, sore throat and inadequate measles eruption (Anonymous 1989).

Arisaema intermedium Blume var. intermedium

Common name: Sarpa Makai **Family:** Araceae
Chromosome number(s): 2n=28

Description: A hardy rhizomatous rooted perennial herb with arum-like flowers. Corm globose, 2 cm in diameter. Leaf solitary, trifoliate; median leaflet ovate, 4 cm broad; lateral leaflet lanceolate, oblique at the base, reticulate; petiole 13 cm long, peduncle shorter than petiole. Spathe green, cylindric, 4 cm long, 2 cm in diam., smooth inside. Spadix unisexual; appendage very long, flagelliform, exerted from the tube; anther 4, dehiscent by a horseshoe-shaped slit.

Distribution: WCE; alt. 2200-3000 m.
Himalayas (Kashmir to Sikkim). *A. intermedium* Blume var. *biflagellatum* (H. Hara) H. Hara: W; alt. 3500 m. Himalayas (Nepal, Sikkim).

Part(s) used: Roots and leaves.

Uses: The paste of roots is applied on the ulcer to remove the puss. Decoction of crushed leaves is used for the treatment of fever (Bhattacharjee 1998).

Arisaema tortuosum (Wall.) Schott

Synonym(s): *Arum tortuosum* Wall **Common name(s):** Sarpa makai (Nep.)
Family: Araceae

Description: A perennial herb monoecious or male with spheroidal tubers. Leaves 2, pedatisect, leaflets 7-13, oblong to lanceolate, acuminate, peduncle longer than the leaf. Tube of spathe green, cylindric, to 7.5 cm long and 3.5 cm diam; blade ovate or oblong-ovate, shortly acute. Spadix unisexual, anthers 2-3, longitudinally dehiscent, appendages elongate, sigmoidly curved in the lower part, almost upright on the upper part.

Distribution: WCE; alt. 1300-2900 m
Himalayas (Punjab to Bhutan), NE India (Meghalaya, Manipur), N. Myanmar, W. China.

Part(s) used: Corm, roots and seeds.

Uses: The plant is poisonous. Corm possesses insecticidal and insect repellant properties. Root is used to kill worms which infects cattle. Seeds are given with salt for colic in sheep.

Artemisia indica Willd.

Synonym(s): *Artemisia vulgaris* auct.; *A. grata* Wall.; *A. grata* Besser
Common name(s): Titepati (Nep.); Mug-wort (Eng.)
Family: Compositae
Chromosome number(s): 2n=34

Description: A perennial aromatic pubescent or villous shrub attaining to 2.4 m high with leafy and paniculately branched stem. Leaves ovate, with stipule-like lobes at the base, deeply pinnatisect; lobes entire, toothed or again pinnatisect; upper leaves smaller, 3-fid, or entire, lanceplate. Heads 3-4 mm long, ovoid or subglobose, solitary or 2-3 together, sessile or very shortly-pedicelled in panicled racemes. Outer flowers female, inner bisexual, fertile. Achenes small, oblong ellipsoid.

Distribution: CE; alt. 300-2400 m.
Himalaya, India, Myanmar, Thailand, S. China, Japan.

Parts used: Young shoots and leaves.

Uses: Plant is stomachic, purgative, deobstruent, antispasmodic, anthelmintic, insecticide, and prescribed in infusion and electuary in cases of obstructed menses and hysteria. It cures asthma, itching, anorexia, gastritis, rheumatism, bronchitis, fever, headache, haemorrhage and diarrhea. Externally, it is used in skin diseases and foul ulcers. It is applied to the head of the young children for the prevention of convulsion.

Asparagus recemosus Willd.

Synonym(s): *A. volubilis* Buch.-Ham
Common name(s): Kurilo, Satawari (Nep.); Satawari (Sans.); Asparagus (Eng.)
Family: Liliaceae
Chromosome number(s): 2n=20

Description: A perennial, spinous, foliage, woody climber with tuberous rootstock and angular branches. Leaves linear-subulate, shorter than the spines. Spines suberect or subrecurved, 0.5-0.8 cm long, cladodes 0.4-1.1 cm long, in tufts of 2-6, spreading, falcate, acuminate, channeled beneath. Racemes solitary or in fascicles, 1.5-2 cm long, simple or branched, pedicels slender, 0.3-0.4 cm long, jointed at the middle. Perianth 0.2-0.3 cm across; lobes spreading, white. Anthers minute. Berry 0.5-0.6 cm in diameter (Mall *et al.* 1986).

Distribution: CE. alt. 600-2100 m. Himalayas, India, Malaysia, Australia, Africa.

Parts used: Roots.

Important biochemical constituent(s): Disaccharides.

Uses: Dried roots (about 700 g) are burnt and the fumes are inhaled for curing fever.

Astilbe rivularis Buch.-Ham. ex D. Don

Common name(s): Thulo okhati, Budho okhati (Nep.) **Family:** Saxifragaceae
Chromosome number(s): 2n=14, 28

Description: An erect herb to 2 m tall. Leaves irregularly bipinnate, lower pinnules usually with 3 leaflets. Leaflets ovate, often cordate to 9 cm long and 4.5 cm broad acuminate, sharply and unequally toothed, upper surface glabrous with few hairs, lower minutely bristly along the midrib and nerves. Inflorescence terminal panicle, 75 cm long, rachis woolly pubescent and brown villous, pale yellow. Flowers small, 0.2 cm across, shortly pedicelled, pedicel pinkish with white hairs. Fruit a capsule, 0.5 cm long, black with numerous minute seeds.

Distribution: WCE; alt. 2000-3600 m.

Himalayas (Kashmir to Bhutan),NE India, Thailand, N. Indo-China, W. China.

Part(s) used: Rhizomes.

Uses: It is used as a tonic. Powder of rhizome is given in pre- and postpregnancy.

Azadirachta indica A. Juss.

Common name(s): Neem Nep.); Nimba (Sans.)
Family: Meliaceae
Chromosome number(s): 2n=28, 30

Distribution: CE; alt. 900 m. Himalayas, India, widely cultivated.

Description: An evergreen tree of 12 m high. Leaves to 30 cm long, pinnate, leaflets 7 cm, lanceolate, acuminate, serrated, 12-17 in number. Flowers small, white, with pleasant odor, in loose clusters. Fruits round, oblong, 1.7 cm diameter, greenish yellow when ripe.

Part(s) used: Every part of the tree and the oil are used in medicine. Bark yields a clear amber colored gum used in medicines.

Important biochemical constituent(s): Nimbin molecule with an acetoxy, a lactone, an ester, methoxy and an aldehyde group.

Uses: Bark of the root and stem and leaves have antibiotic activity. Green twigs are used to brush teeth and cures toothache, bad breath and gum diseases. Preparation of the bark is a bitter tonic and astringent and used to treat fever and skin diseases. Leaves are also bitter, used in skin diseases and boils. Decoction of fresh leaves cures fever and removes phelgm from bronchial tubes. Smoke from burnt leaves repels mosquitoes. Seeds yield an acrid, bitter oil of deep yellow having disagreeable odor. Seeds yield 30-40% oil which is used as anthelmintic and antiseptic. Sap of neem cures leprosy. Ointments made from this plant is used for healing ulcers and wounds (Bhattacharjee 1998).

Baliospermum montanum (Willd.) Müll. Arg.

Synonym(s): *Baliospermum axillare* Blume **Common name(s):** Ajayaphal (Nep.); Danti (Sans.) **Family:** Euphorbiaceae
Chromosome number(s): 2n=44

Description: An erect dioecious undershrub. Fruit a 3-lobed capsule with mottled seeds and oily endosperm; seeds resembling those of castor plants and measuring 0.8 - 1.4 cm long and 0.5 - 0.8 cm broad; seed coat hard.

Distribution: WCE; alt. 300-910 m.
Himalayas (Uttar Pradesh to Bhutan), India, Indo-China, Malaysia.

Part(s) used: Roots and the seeds.

Important biochemical constituent(s): Root -5 new phorbol esters belonging to terpine hydrocarbon and lignine skeleton.

Uses: The root is a mild cathartic. Seeds act as a strong purgative. In overdoses they are highly poisonous. Used as stimulant and rubifacient. Oil of seeds hydrogogue, cathartic and externally used in rheumatism.

Bauhinia purpurea L.

Synonym(s): *B. triandra* Roxb.
Common name(s): Tanki, Koiralo (Nep.) **Family:** Leguminosae
Chromosome number(s): 2n=26, 28

Description: A deciduous tree to 6 m. tall with smooth dark brown bark sometimes having silvery patches. Leaves with characteristic deep cleft (to half) at the apex separating two rounded usually blunt lobes. Inflorescence terminal or axillary racemes having few flowers in clusters. Flowers large, fragrant, pink-purple, on leafless branches. Pod glabrous, to 14.5 cm long, dehiscent.

Distribution: WCE; alt. 300-1600 m.
Tropical Himalayas (Kashmir to Bhutan), India, SE Asia, W. & S. China.

Part(s) used: Bark.

Uses: Bark is used in diarrhea and dysentery. It is also used as astringent.

Bauhinia vahlii Wight & Arn.

Synonym(s): *Bauhinia racemosa* Vahl
Common name(s): Bhorlo (Nep.); Mhata, Mhalavata, Kusalapte (Npbh.); Murva (Sans.); Camel's foot climber (Eng.)
Family: Leguminosae
Chromosome number(s): 2n=28

Description: A large evergreen climber with rusty-brown-haired leaves and shoots, and with stems to 1.6 m in diameter. Leaves variable in size, rounded to 46 cm across with cordate base and cleft to about one third into 2 lobes with rounded apices, dark green above, downy beneath. Flowers creamy, 3.5-5 cm across, in dense-stalked, flat-topped terminal cluster. Pod woody, to 30 cm long and 5 cm broad, with rusty velvet hairs. Seed flat, polished.

Distribution: WCE; alt. 200-1300 m.
Tropical Himalayas (Kashmir to Sikkim), India.

Part(s) used: Seeds.

Uses: Seeds are used as aphrodisiac and tonic.

Bauhinia variegata L.

Synonym(s): *Bauhinia candida* Alton
Common name(s): Koiralo (Nep.); Kunahbun (Npbh.); Kancanara, Kovidara (Sans.); Mountain ebony (Eng.)
Family: Leguminosae
Chromosome number(s): 2n=28

Description: A medium-sized deciduous tree with ashy to dark brown bark. Leaves cleft (to a third) at the apex into 2 rounded lobes. Flowers fragrant, 5-6 cm, petals all white or 4 petals pale purple, and the fifth darker with purple veins.

Distribution: WCE; alt. 150-1900 m.
Himalayas (Swat to Bhutan), India, Myanmar, China.

Part(s) used: Bark, leaf, young shoot and flower.

Uses: About 25 ml of decoction of the bark is administered twice a day for two weeks in tubercular lymphadenia. The bark contains tannin. Flower powder is believed to cure haemorrhage.

Belamcanda chinensis (L.) Redoute

Common name(s): Blackberry lily , leopard flower, dwarf-tiger lily (Eng.).
Family: Iridaceae
Chromosome number(s): 2n=32

Description: A perennial erect herb with flat, divaricate and brownish rhizomes bearing scars of the resinous stem on the surface. Rootstock creeping, leafy. Leaves ensiform, with short sheaths, 30 cm long and 2-3 cm broad. Spathes several-fold, subscariose, bracts scariose. Flowers pedicelled. Perianth rotate in 6 segments, oblong, orange, red spotted. Capsule obovoid, membranous. Seeds subglobose.

Distribution: WC; alt. 1300-2300 m.
India, Himalayas, Malaya, SE Asia (as an escape from cultivation), China, Japan.

Part(s) used: Rhizomes and roots

Uses: Rhizomes and roots are given in sputum, retropharyngeal abscess, sore throat tracheitis and tonsillitis.

Berberis aristata DC. var. aristata

Common name(s): Chutro, Rasanjan (Nep.); Maripyasi (Npbh.); Daruharidra (Sans.); Berberry (Eng.)
Family: Barberidaceae
Chromosome number(s): 2n=28

Description: A robust shrub, about 2 m tall with spineless stem and arching pale yellow branches. Leaves glossy green, often spineless, usually elliptic. Flowers yellow, about 6 mm long, numerous in a short-stalked cluster. Fruit dark red, blue-purple when ripe.

Distribution: WC; alt. 1800-3000 m. India, Himalayas (Nepal)

Part(s) used: Root bark and wood.

Important biochemical constituent(s): Berberine.

Uses: Alterative, astringent, antiperiodic, deobstruent, used in skin diseases, menorrhagia, diarrhea and jaundice. Decoction of root is useful in controlling fever. Root bark is used externally in eye diseases.

Berberis asiatica Roxb. ex DC.

Common name(s): Chutro, Rasanjan (Nep.); Maripyasi (Npbh.); Berberry (Eng.)
Family: Barberidaceae
Chromosome number(s): 2n=28

Description: A branched shrub to 3 m tall with yellow wood. Stem with spines of 1-1.5 cm long. Leaves thick, rigid, evergreen with 2-5 spiny teeth, shining dark green above and grayish beneath, obovate to elliptic to 7.5 cm long. Inflorescence corymbose raceme to 3 cm long. Flowers small, yellow. Fruit a berry, black or reddish deep purple.

Distribution: WCE; alt. 1200-2500 m.
Himalayas (Uttar Pradesh to Bhutan), NE India, China (Yunnan)
Part(s) used: Root.
Important biochemical constituent(s): Berberine.
Uses: The extract of root is used in the eye infection. Root decoction is used in body inflammation and stomachache.

Bergenia ciliata forma ligulata Yeo

Synonym(s): *Bergenia ligulata* (Wall.) Engl.
Common name(s): Pashanved (Nep.)
Family: Saxifrageceae
Chromosome number(s): 2n=34

Description: A perennial rhizomatous herb with very stout and woody rootstock. Leaves orbicular or obovate, entire, ciliate, with cordate base and rotundate apex, to 20 cm long and 15 cm broad, 4-6 pairs of nerves, glabrous on both surfaces. Inflorescence helicoid cyme bifurcating at the tip, to 30 cm long. Flowers pink or pinkish-white. Fruit a capsule, 2 mm long.

Distribution: WC; alt. 1600-3200 m.
Afghanistan, Himalayas (Kashmir to Bhutan), China (Xizang), NE India.

Conservation status: Commercially threatened (IUCN Category). **Part(s) used:** Rootstock.

Uses: Rhizomes and roots are bitter, astringent, diuretic, demulcent, aphrodisiac; also useful in fever, diarrhea, pulmonary afflictions and renal and muscular calculus. They are also used after childbirth, and applied to boils and ophthalmia.

Betula utilis D.Don

Synonym(s): *Betula bhojpattra* Lindl.
Common name(s): Bhojpatra (Nep.); Bhurjapatra (Sans.); Himalayan Birch (Eng.)
Family: Betulaceae
Chromosome number(s): 2n=28, 56

Description: A tree to 15 m high with smooth, paper-like bark, peeling up in thin sheets, outer layer pinkish white and inner layer pink. Leaves to 7.6 cm, long, alternate, petiolate, ovate, serrated and long pointed. Male flowers in pendulous spike and female flower in erect or pendulous spike.

Distribution: WCE; alt. 2700-4300 m Himalayas (Nepal to Bhutan), W. China.

Part(s) used: Bark.

Important biochemical constituent(s): Essential oil and betulin (Anonymous 1994).

Uses: Infusion of bark is antiseptic and carminative. Also used in hysteria.

Boerrhavia diffusa L.

Synonym(s): *Boerrhavia repens* L.
Common name(s): Punarnava (Nep.); Punarnava (Sans.)
Family: Nyctaginaceae
Chromosome number(s): 2n=26, 52, *c.*116

Description: A deep-rooted perennial spreading herb. Leaves two in each node - one smaller than the other, upper surface green and the lower surface whitish, base cordate. Flowers in small clusters on long axillary stalks, very small, reddish in color. Fruits glandular with fine ridges.

Distribution: WCE; alt. 300-1200 m.
Tropical and subtropical regions of the Old World.

Part(s) used: Whole plant.

Important biochemical constituent(s): An alkaloid punarnavine, sulfates, chlorides and traces of nitrate and chlorates have been isolated from the plant. Roots contain Hentriacontane, β-sitosterol and ursolic acid (Anonymous 1994).
Uses: Diuretic and laxative. It is also used to treat asthma, dropsy, jaundice, intestinal inflammation and gonorrhoea. Root powder is used in eye diseases.

Bombax ceiba L.

Synonym(s): *Bombax malabaricum* DC.
Common name(s): Simal (Nep.); Kantakadruma (Sans.) **Family:** Bombacaceae
Chromosome number(s): 2n=72, 92, 96

Description: A very large tree attaining the height of 45 m and the girth of 12 m with horizontal branches arranged in whorls of 5-7. Stem with sharp conical prickles. Leaves with 5-7 lanceolate leaflets, each leaflet 10-20 cm long. Flowers large, scarlet, occasionally white; petals 5-7.5 cm long and tomentose; anthers long and twisted. Capsule oblong, hard, woody, 10-12 cm long. Seeds glabrous, embedded in silky wool.

Distribution: CE; alt. 200-900 m.
Tropical Himalayas, India to W. China, Malaysia.

Part(s) used: Root, gum

Uses: The resin powder is used in diarrhea.

Brachycorythis obcordata (Lindl.) Summerh.

Common name(s): Gamdol (Nep.) **Family:** Orchidaceae
Chromosome number(s): 2n=42
Description: A terrestrial orchid to 25 cm tall. Tubers globose or oblong to 3 cm long and 1.5 cm across. Stem with a few loose leaf-sheaths on its lower part. Leaves oblong-lanceolate or elliptic-lanceolate to 5 cm long and 2 cm broad, blunt or subacute. Inflorescence a spike, to 8 cm long. Flowers 1-2.2 cm across, pink or pale purple.
Distribution: WCE; alt. 1000-2000 m. Himalayas (Himachal Pradesh to Bhutan).
Part(s) used: Tuber.
Uses: Tubers are expectorant, astringent and nutritious.

Butea buteiformis (Voigt) Grierson

Synonym(s): *Butea minor* Buch.-Ham. ex Baker
Common name(s): Palabi (Npbh.)
Family: Leguminosae
Chromosome number(s): 2n=18

Dr. Kamal K. Joshi and Prof. Sanu Devi Joshi

Description: A shrub to 1.5 m tall with silky-haired young branches. Leaves very large trifoliate; leaflets broadly ovate, to 38 cm, terminal leaflet long-stalked and the lateral ones asymmetrical, leathery, finely silky-hairy beneath. Flowers numerous, dark red in long terminal and axillary spike-like clusters. Pod densely hairy, 5-8 cm long and 2.5-3 cm broad, flat, slightly biconcave, whitish-gray, usually single-seeded.

Distribution: WCE; alt. 300-2000 m. Nepal, NE India.

Part(s) used: Seed.

Uses: The seed is bitter and anthelmintic.

Butea monosperma (Lam.) Kuntze

Synonym(s): *Butea frondosa* J. König ex Roxb.
Common name(s): Panlas (Nep.)
Family: Leguminosae
Chromosome number(s): 2n=18, 18+f, 32

Description: A medium-sized deciduous tree, with twisted and gnarled trunk. Bark fibrous and light brown. Leaves rough in texture, pinnately trifoliate; leaflets coriaceous, ovate, 10-15 cm long. Flowers bright flaming scarlet orange with black calyces; petals pubescent, 5-7 cm long. Ripe pods light and disseminated far and wide by the wind.

Distribution: WCE; alt. 150-1200 m.
Tropical Himalayas, India, Sri Lanka, SE Asia, Malaysia.

Conservation status: Endangered (IUCN Category). **Part(s) used:** Bark, leaves, gum, flowers and seeds.

Important biochemical constituent(s): Flowers contain flavonoid glucosides butrin, isobutrin, coreopsin, isocoreopsin and sulfurein (Anonymous 1994).

Uses: The leaves and flowers are astringent, depurative, diuretic and aphrodisiac. These are used against boils and pimples. These are also taken internally in flatulent colic, worms and piles. Gum contains tannins and is used during diarrhea. Mixed decoction of flowers and seeds is used as against tapeworms.

Callicarpa macrophylla Vahl

Synonym(s): *Callicarpa incana* Roxb.
Common name(s): Priyangu, Dahikamlo (Nep.); Priyangu, Gandhaphali (Sans.)
Family: Verbenaceae
Chromosome number(s): 2n=34

Description: Evergreen shrub to 3 m tall. Twigs, leaf-stalks, flower-stalks and undersurfaces of leaves densely woolly-hairy. Leaves opposite, ovate to oblong lanceolate, long-pointed, rounded-toothed, stalked. Flowers mauvishpink in rounded many-flowered branching axillary clusters. Calyx companulate, minutely 4-lobed. Corolla 2.5 mm long, with cylindric tube and 4 spreading lobes. Stamens exserted. Fruit white, globose, succulent.

Distribution: WCE; alt. 300-1500 m.
Himalayas (Kashmir to Bhutan), India, Myanmar, S. China, Indo-China.

Part(s) used: Flowers and fruits.

Uses: Powder of flowers and fruits (3-6 g) or unripe fruits (10-20 g) are given 2-3 times a day to treat mouth ulcer, burning sensation of hands and legs and hyperacidity in Ayurvedic treatment.

Calotropis procera (Aiton) Dryand.

Common name(s): Alkarka, Arka (Sans.); Aank (Nep.); Swallow-wort (Eng.)
Family: Asclepiadaceae
Chromosome number(s): 2n=22, 26

Description: A perennial undershrub 2 m tall with white latex. The bark and the leaves are grayish in color. Leaves 10-13 cm long and 4-7.5 cm broad, obovate, undersurface pubescent. Flowers white, purple, or violet in clusters, sweet scented. Fruits to 4 cm long, with soft white floss and numerous black seeds.

Distribution: C; alt. 170-600 m.
Tropical Africa, W. Asia, tropical Himalayas, India.

Part(s) used: Root and root bark, leaves and latex from the leaves and flowers.

Uses: The paste prepared from the root charcoal by mixing it with some bland oil is applied over skin diseases caused by syphilis, leprosy etc. The root bark is cholagogue, diaphoretic, emetic, alterative and diuretic.

The leaves are useful in dropsy and enlargement of the abdominal viscera. Dry leaves are either smoked or the smoke of the burning leaves is inhaled for the cure of asthma and cough. A decoction of leaves is used as wormicide. The leaf juice is applied to skin diseases.

The latex is used in leprosy, dropsy, rheumatism etc. As an abortifacient it is either taken internally or applied to the mouth of the uterus.

The flowers are used as a tonic and stomachic. They are also used during indigestion.

Caltha palustris var. himalensis (D.Don) Mukerjee

Synonym(s): *Caltha himalensis* D. Don; *C. paniculata* Wall. **Common name(s):** Ta mik chewa (Am.); Marsh marigold (Eng.) **Family:** Ranunculaceae
Chromosome number(s): 2n=32

Description: A perennial herb attaining a height of about 90 cm. Leaves, glabrous, shining, rounded or cordate, mostly basal, long-stalked, blades 3-15 cm across, finely toothed. Flowers few usually white or yellow cup-shaped, 2.5-4 cm across with 5-8 ovate petals and numerous stamens and carpels. Fruit a cluster of beaked follicles each with *c.* 1 cm.

Distribution: WCE; alt. 2400-4200 m. Himalayas (Uttar Pradesh to Bhutan).

Part(s) used: Leaves.

Uses: Leaves are vesicant and very bitter and used as febrifuges (Bhattacharjee 1998).

Canna edulis Ker-Gawl.

Common name(s): Queensland arrowroot (Eng.) **Family:** Cannaceae
Chromosome number(s): 2n=18, 27

Description: A perennial herb of 3 m tall with rhizomatous roots. Flowers bright red.

Distribution: E; alt. 1200-1800 m.
Cultivated; Native of S. America and W. Indies.

Part(s) used: Rhizome

Uses: Rhizome is used to make emollient poultices for abscesses and tumors (Bhattacharjee 1998).

Cannabis sativa L.

Common name(s): Bhang, Ganja (Nep.); Gaji (Npbh.); Bhanga, Vijaya (Sans.); Hemp (Eng.)
Family: Cannabaceae
Chromosome number(s): 2n=20

Description: Erect slender herb up to 2 m tall with ridged and pubescent stem. Stipules filiform; leaves 3-11-foliate, upper ones one-foliate; leaflets sessile, narrowly lanceolate to linear-lanceolate, long, acuminate, serrate; upper surface of the leaflets scabrid, lower surface adpressed and pubescent. Anthers yellow. Female flowers: perianth 0, outer surface of the enveloping bract clothed with glandular hairs. Stigmas prominent, thread-like, 1/2 cm long. Fruit shining, yellow to brown (Malla *et al.* 1986).

Distribution: WCE; alt. 200-2700 m.
Probably native of C. Asia; widely cultivated and naturalized in most temperate and tropical areas of the world.

It seems likely that all the Nepalese materials of *C. sativa* L. belongs to *C. sativa* var. *kafiristanica* (Vavilov) Small & Cronquist. Detailed chemical analysis is necessary before more definite determinations can be made (Press *et al.* 2000).

Part(s) used: Dried branches and the resinous exudate from the inflorescence.

Important biochemical constituent(s): Resin, volatile oil, cannabinoids, cannabispirams and alkaloids (Rajbhandari *et al.* 1995).

Uses: Tonic, intoxicant, stomachic, antispasmodic, analgesic, narcotic, sedative and anodyne. Resinous exudate is used as hashis. It is also given in diarrhea, dysentery and cholera (Malla *et al.* 1997).

Capparis spinosa L.

Common name(s): Ban bhendo (Dolp.); Capers or caper bush (Eng.)
Family: Capparaceae
Chromosome number(s): 2n=24, 38

Description: An evergreen low trailing or prostrate spiny shrub with heavily clothed foliage. Leaves variable, ovate to elliptic, 1-4 cm, mostly spine-tipped; stipules of 2 hooked brown or yellow spines. Flowers whitish to pinkish, solitary, 3-8 cm across, long-stalked, borne in the axils of the younger leaves; petals 4, obovate, 2-5 cm, stamens purplish, much longer. Fruits fleshy, oblong-ellipsoid, 2-5 cm, long-stalked with red flesh and brown seeds.

Distribution: *C. spinosa* L. var. *himalayansis:* W; alt. 2100 m. Himalayas (Kashmir to Nepal).
C. spinosa L. var *spinosa:* ECW; alt. 200-2400 m.
S. Europe, Africa, C. Asia, Himalayas, India.

Part(s) used: Root bark, leaves, flower buds and fruits.

Important biochemical constituent(s): Rutin pentosans, rutic acid, pectic acid, volatile emetic constituents and saponin.

Uses: The root bark is slightly bitter and tart. It is aperative, diuretic, resolvent, and tonic. It facilitates digestion and stimulates appetite. It is used to treat rheumatism, paralysis, toothache, tubercular glands and afflictions of liver and spleen. Bruised leaves are used as poultice in gout. Flower buds and fruits are useful in scurvy. Pickled flower buds are sold as "capers".

Carum carvi L.

Common name(s): Bhote jeera (Dolp.); Krishnajiraka, Sushavi (Sans); Caraway (Eng.)
Family: Umbelliferae
Chromosome number(s): 2n=20, 22

Description: A biennial herb of 60 cm tall with fleshy roots. Leaves compound, pinnate, divided into very narrow segments. Flowers small, white and borne in flat compound umbels. Dry fruits brown in color and with pleasant odor. Seeds brown and hard.

Distribution: WC; alt. 2500-5100 m.
Karakoram, Himalayas (Kashmir, Nepal, Bhutan), China (Xizang).

Part(s) used: Fruits.

Important biochemical constituent(s): Carvone, d-limonene, carveol, diacetyl furfural, methyl alcohol, acetic aldehyde, thiamine, riboflavin, niacin and ascorbic acid (Bhattacharjee 1998).

Uses: Widely used as a spice for culinary purposes and for flavoring bakeries. Its oil is a strong carminative, stimulant and aromatic. It is also used as poultices in the swelling of breast and testicles. It is a mild stomachic and sometimes used in flatulent colic. Carvone is anthelmintic (Bhattacharjee 1998).

Cassia fistula L.

Synonym(s): *Cassia rhombifolia* Roxb.
Common name(s): Rajbriksha (Nep.); Aragvadh, Prapunnada, Suvarnaka (Sans.); Purging cassia or Indian laburnum (Eng.)
Family: Leguminosae
Chromosome number(s): 2n=24, 26, 28

Description: A small nearly evergreen tree with greenish-gray bark and large pinnate leaves. Leaves with 3-8 pairs of opposite ovate to elliptic leaflets 5-15 cm, pale silvery-haired below when young. Flowers bright yellow, 5 cm across, in drooping lax clusters of 30-60 cm long. Petals 5 obovate, veined. The 3 lowest stamens longer and curled, with large anthers, 4 or 6 lateral stamens smaller, straight; 1 or 3 upper stamens short, incurved and sterile. Pod 40-60 cm, rounded in section, brown to glossy black, not splitting, many-seeded.

Distribution: WCE; alt. 150-1400 m.
Widely cultivated in Africa, W. Asia, Himalayas, India, SE Asia, Malaysia, China, Polynesia; probably a native of E. India, Myanmar, Malay islands.

Part(s) used: Bark and fruit.

Important biochemical constituent(s): The bark of the tree is rich in tannin. The pulp contains sugar, tartaric acid, melic acid, oxalic acid and cathartic acid.

Uses: "Cassia pulp" is a well known laxative. Cathartic acid is considered the purgative principle of the drug (Bhattacharjee 1998).

Cassia tora L.

Common name(s): Chakramarda, Dadamari, Prapunnada (Sans.); Sickle senna, fetid senna, ring-worm plant (Eng.)
Family: Leguminosae
Chromosome number(s): 2n=24, 26, 28, 52

Description: An annual herb to 90 cm tall. Leaves 8-12 cm long; leaflets 6, oboval or obovate-oblong, obtuse, base attenuate, 3-5 cm long 1.5-2.5 cm broad, glabrous. Flowers grouped 1-3 in the leaf axils. Fruit a linear pod, 12-14 cm long and 0.4 cm wide. Seeds oblong, 3-5 mm long and 2-3 mm diameter pointed at one end, round or truncate at the other, deep brown, smooth and glossy.

Distribution: WCE; alt. 450-1300 m. Tropics, probably of S. American origin.

Part(s) used: Seeds.

Uses: Seeds are bitter and mucilaginous. The seed paste is used in the treatment of ringworm and itch. About 5 g of crushed seeds is taken with water in cough (Bhattacharjee 1998). Seeds (9-15 g) are given to treat acute conjunctivitis, corneal ulcer, night blindness, and glaucoma. It is also useful in dizziness due to hypertension, habitual constipation and infant malabsorption and malnutrition (Anonymous1989).

Cedrus deodara (Roxb. ex D. Don) G. Don

Synonym(s): *Cedrus libani* var. *deodara* (Roxb. ex D. Don) Hook. f. **Common name(s):** Devdaru (Nep.); Devdaru (Sans.); Himalayan cedar (Eng.) **Family:** Araucariaceae
Chromosome number(s): 2n=24

Description: A large robust tree to 80 m, with a girth to 10-12 m. Branches spreading with drooping branchlets. Bark smooth and with vertical fissures. Leaves solitary or in dense clusters, dark green or sometimes bluish-green, rigid, leathery, 3-sided, sharply pointed, 2.5-4 cm long. Male cones cylindrical, 5-12 cm, erect; female cones ovoid-cylindrical, erect, large, 10-13 cm long and 8-10 cm broad, with numerous thin scales; young cones bluish-purple.

Distribution: WC; alt. 2000-2500 m.

Afghanistan, Himalayas (Kashmir to Nepal).

Part(s) used: Dried bark.
Important biochemical constituent(s): Deodarin, dihydroflavonol, taxifolin and other flavones.
Uses: The dried bark is astringent and used in fevers, diarrhea and dysentery.

Celosia argentea L. var. argentea

Common name(s): Sitavarka, Vitunna (Sans); Feather cockscomb, quail grass (Eng.).
Family: Amaranthaceae
Chromosome number(s): 2n=72

Description: An annual erect herb, 0.3-1.0 m tall, glabrous, more or less branching. Leaves variable, linear or lanceolate, 8-10 cm long and 2-4 cm wide, acuminate, entire, tapering at the base. Flowers pinkish gradually turn into white, crowded and imbricate, in close imbricate, cylindrical, terminal spikes, to15 cm long and 2.5 cm in diameter. Perianth 8 mm long or more. Capsules 3-4 mm long, ellipsoid, tapering at the apex in the style. Seeds black, subreniform, shining, flat, small.

Distribution: WCE; alt. 1500-1600 m.

Dr. Kamal K. Joshi and Prof. Sanu Devi Joshi

Cosmopolitan. *C. argentea* var. *cristata (L.)* Kuntze: C. Tropical garden escape.

Part(s) used: Seeds.

Uses: The seed is bitter and used to cure acute conjunctivitis, keratitis, chronic uveitis and dizziness due to hypertension (Anonymous 1989).

Celtis australis L.

Common name(s): Nettle tree (Eng.)
Family: Ulmaceae
Chromosome number(s): 2n=40

Description: A moderate to large-sized deciduous tree with drooping branches. Bark smooth, bluish-gray, often with horizontal wrinkles and small rounded swellings. Leaves simple, alternate, oval, mature ones tough and leathery, short-stalked, prominently 3-veined at the base, toothed. Flowers small, pale yellow, or greenish, appear with the new leaves, male, female or both in the same plant, male and hermaphrodite borne on short stalks in tuffs at the base of the shoot, female flowers long-stalked, borne at the leaf axils. Fruits fleshy, usually round, *c.* 1 cm diam., variable considerably in shape and size.

Distribution: C; alt. 1300-2200 m. Himalayas (Kashmir to Nepal), India.

Part(s) used: Fruits.

Uses: Cures stomach disorders (Adrian and Storrs 1990).

Centella asiatica (L.) Urb.

Synonym(s): *Hydrocotyl asiatica* L.
Common name(s): Ghodtapre (Nep.); Kholcha ghayen (Npbh.); Mandukaparni (Sans.); Water pennywort, Indian pennywort (Eng.)
Family: Umbelliferae
Chromosome number(s): 2n=18, 22, 33

Description: A prostrate herb with orbicular-reniform simple leaves. Leaves 5.0-7.5 cm, entire, crenate or lobulate. Inflorescence an umbel with very short peduncle and 3-6 sessile flowers.

Distribution: WCE. alt. 500-2100 m.
Nepal. Widespread in tropical and subtropical regions throughout the world.

Part(s) used: Whole plant.

Important biochemical constituent(s): Vallerin, a mixture of tri-terpenoid glycosides (mostly asiaticoside), traces of alkaloid, volatile oil and pectin.

Uses: The plant is used as a tonic and alterative for skin diseases, leprosy and nerves. It is also used as blood purifier, diuretic and insecticide. It is given in indigestion and nervousness. Leaf stalk is used in toothache.

Chenopodium album L.

Common name(s): Bethe (Nep., Dolp.); Pigweed (Eng.)
Family: Chenopodiaceae
Chromosome number(s): 2n=18, 36, 42, 54

Description: An annual herb to 1 m tall, usually grown as a weed in wheat field, with often purple-tinged stem and leaves. Leaves 2.5-6 cm long, the lower ovate or oblong, with wavy toothed margin, stalked, the upper often narrow, entire. Flowers small, green in rounded clusters, borne in slender spikes or forming a lax leafy-branched terminal inflorescence. Perianth-segments keeled, covering the fruit. Seeds smooth.

Distribution: WC; alt. 2000-4000 m. Cosmopolitan weed.

Part(s) used: Leaves and shoots.

Important biochemical constituent(s): Carotene, linolenic acid, vitamin C and iron have been isolated from the plant.

Uses: Mild laxative. Leaves and shoots are used to drink as tea which relieve pain in stomach.

Dr. Kamal K. Joshi and Prof. Sanu Devi Joshi

Cimicifuga foetida L.

Synonym(s): *Cimicifuga frigida* royle
Common name(s): Gya tsi goepa (Am.); Skunk bugbane, stinking bugbane (Eng.)
Family: Ranunculaceae
Chromosome number(s): 2n=16, 32

Description: A perennial, more or less pubescent herb with dark-brown irregular rhizome measuring 10-20 cm long and 2-4 cm in diameter. Stem erect, leafy, branched. Leaves pinnately compound; leaflets ovate or lanceolate, deeply and sharply toothed, terminal leaflet 3-lobed. Flowers nearly regular, about 6 mm in diameter, white, crowned by short or long racemes, solitary in the axils of the upper leaves and combined in a terminal panicle, sometimes large and spreading.
Distribution: WCE; alt. 3000-4000 m.
Himalayas (Kashmir, Uttar Pradesh to Bhutan), N. Myanmar, W. & N. China, Mongolia, Korea.
Part(s) used: Rhizomes
Uses: Rhizome is bitterish and astringent. It is given in the treatment of toothache, headache, sore throat and also used in chronic diarrhea (Anonymous 1989).

Cinnamomum camphora (L.) J. Presl

Common name(s): Karpur (Nep.); Kapoo (Npbh.); Karpura (Sans.); Camphor (Eng.)
Family: Lauraceae
Chromosome number(s): 2n=12, 24

Description: A medium-sized tree of to 12 m high. Leaves and twigs have a smell of camphor. Leaves smooth, shining, whitish beneath, to 12 cm long. Flowers yellow.

Distribution: C; alt. 1300-1500 m.
Native of Japan; widely cultivated elsewhere.

Part(s) used: Twigs and leaves.

Important biochemical constituent(s): Essential oil and tannin.

Uses: Source of camphor. Aromatic and stimulant. Oil is used to treat diarrhea, rheumatism and muscular pains. Useful in bronchitis and pneumonia. Also stimulates uterus, menstruation and uterine hemorrhages (Bhattacharjee 1998).

Cinnamomum glaucescens (Nees) Hand.-Mazz.

Common name(s): Sugandhakokila (Nep.)
Family: Lauraceae

Description: An evergreen tree about 15 m high with spreading branches. Leaves variable, often glaucous beneath. Flowers small, 2.5 mm across. Fruits globose, reddish brown when ripe (Malla *et al.* 1996).

Distribution: WE; alt. 2000-2500 m. Himalayas (Nepal to Bhutan), India.

Conservation status: Banned for export outside the country (under the Forest Act 1993).

Part(s) used: Fruits and wood.

Important biochemical constituent(s): D-camphor and essential oils.

Uses: The pericarp of the fruit yields essential oils used in perfumery, incense sticks, soap and toiletries, etc. Wood contains D-camphor and is said to be a good substitute for Sassafras.

Cinnamomum tamala (Buch.-Ham.) Nees & Eberm.

Synonym(s): *Cinnamomum cassia* auct.
Common name(s): Dalchini/Tejpat (Nep.); DalchiniTejpat (Npbh.); Tamalaka/ Tejapatra (Sans.); Indian cassia lignea (Eng.)
Family: Lauraceae
Chromosome number(s): 2n=24

Description: A small evergreen tree to 8 m tall. Leaves bright pink when young in spring, aromatic when crushed, leathery, short-stalked, ovate-lanceolate, long-pointed, entire, 10-15 cm long, glaucous beneath, with 3 conspicuous nearly parallel veins arising from near the base, the leaf-tip often curved. Flowers pale yellow in terminal and axillary-branched clusters about as long as the leaves. Perianth 7 mm long, with oblong lobes and a short tube, silky-haired; fertile stamens 9; ovary hairy, with slender style. Fruit black, succulent, ovoid 12 mm, seated on the somewhat enlarged perianth-tube.

Distribution: WCE; alt. 450-2000 m.
Himalayas (Kashmir to Bhutan), NE India (Assam, Meghalaya).
Part(s) used: Leaves.
Important biochemical constituent(s): Essential oil with major components viz., d-a-phellandene, eugenol and cinnamic acid.
Uses: Leaves are regarded as carminative and spice. It is also used to control diarrhea and colic pain.

Cissampelos pareira Linn.

Common name(s): Gujargano, Patha, Batule pat (Nep.); Ambastha, Laghupatha (Sans.); False Pareira (Eng.)
Family: Menispermaceae
Chromosome number(s): 2n=24

Description: A slender woody climber with short stem and long herbaceous tomentose branches. Leaves to 8 cm in diameter, petioled, orbicular to reniform, usually peltate or cordate at the base, apex obtusely mucronate. Inflorescence of male flowers axillary cyme, 1-3 cm long, and that of female flowers axillary raceme. Flowers small, green; male with minute bracts, slender, pubescent, tomentose or hirsutate peduncle and female with large bracts, reniform or orbicular, sometimes petioled, pedicels very short, stigma black, bifurcated. Fruit yellow, drupe, 0.4 cm diameter, subglobose.

Distribution: WCE; alt. 150-2200 m. Pantropical.

Part(s) used: Whole plant.

Important biochemical constituent(s): Alkaloids such as cycleanine, hayatidine, hayatinin and hayatin have been found in the leaves (Anonymous 1994).

Uses: The root decoction is used to control diarrhea and urinary trouble. The plant possesses anthelmintic, antihistaminic and antipyretic properties. It is also used as cardio tonic and stomachic (Bhattacharjee 1998).

Citrus limon (L.) Burm. f.

Synonym(s): *Citrus medica* subsp. *Limon* L.; *C. medica* var. *limonum* Hook. f.
Common name(s): Kagati (Nep.), Kagati (Npbh.), Bijapura (Sans.), Lemon (Eng.)
Family: Rutaceae
Chromosome number(s): 2n=18, 36

Description: A small tree to 7 m tall, with short spines and white flowers about 2.5 cm across. Fruits small, light yellow when ripe, oval, with blunt ends.

Distribution: C; alt. 1600 m.
India, Nepal, China. Widely cultivated throughout the subtropics.

Part(s) used: Peel of the fruit

Important biochemical constituent(s): Essential oil from the peel of the fruit contains d-limonene, terpinene and cadinene.

Uses: The oil is carminative and flavoring agent. It is also used in perfumery and toilet soaps

Claviceps purpurea (Fr.) Tul

Common name(s): Ergot (Eng.)
Family: Clavicipitaceae

Description: A parasitic fungus belonging to the class ascomycetes grows on the heads of rye (Secale cereale L.) by infecting the flowers with ascospores through the stigma. It forms long horn-like dark purple or nearly black hard fruit-like bodies called sclerotia. The sclerotia takes the place of the seed and arrests the development of the seed of the host plant. It has a fishy odor.

Part(s) used: Ergot.

Important biochemical constituent(s): More than 50 different alkaloids have been identified from the ergot. The primary constituents of ergot sclerotia are a group of indole alkaloids with a tetracyclic ring system called ergoline. The important alkaloids are ergonovine, ergotanine, ergocryptine, ergocornine, ergocristine, ergosine and ergovalide. Similarly, ergosterol, fungisterol, clavicepsin, sclererythin, ergochrysin, ergoflavin, and inorganic salts and proteins have also been isolated from the ergot.

Uses: Physiologically active alkaloids stimulates muscular movement of uterus. Ergonovine is effective in controlling haemorrhage after child birth. It is also administered in migraine. Ergotamine with caffeine is used in some migraine cases. It is reported to constrict cranial blood vessels modifying their pulsation. Ergotoxine group of alkaloids is believed to control hypertension and peripheral vascular disorders. Dihydroergotoxine is given in some cases of senility. Bromoergocrystine, a derivative of ergocryptine is used in breast cancer and galactorrhoea.

Clematis napaulensis DC.

Synonym(s): *Clematis cirrhosa* var. *nepalensis (DC.)* Kuntze; *C. montana* D. Don
Common name(s): Imgo nak po (Am); Clematis (Eng.)
Family: Ranunculaceae
Chromosome number(s): 2n=16

Description: An evergreen glabrous climber with spreading stem often forming dense masses, attaining 12 m in height and 2 cm in diameter. Leaves mostly fascicled, trifoliate; leaflets entire or more or less deeply 3-lobed. Flowers greenish white, axillary, fascicled, slender with drooping peduncles, 25-65 cm long. Achenes hairy.

Distribution: WCE; alt. 1200-2500 m.
Himalayas (Punjab, Uttar Pradesh to Bhutan), N. Myanmar, SW China.

Part(s) used: Whole plant.

Important biochemical constituent(s): Anemonin.

Uses: The plant is acrid and poisonous. Leaves are irritating and harmful to skin.

Coelogyne cristata Lindl.

Family: Orchidaceae
Chromosome number(s): 2n=40

Description: An epiphytic herb with oblong pseudobulb. Leaves sessile, lanceolate. Inflorescence drooping raceme. Flowers to 10 in number, large with 3-lobed lip; midlobe somewhat suborbicular with 2 crenulate yellow plates, lateral lobes round, large with yellow fimbriate lamellae, other perianth lobes subequal, obtuse, wide, white.

Distribution: WCE; alt. 1000-2000 m.
Himalayas (Uttar Pradesh to Sikkim), NE India (Meghalaya).

Conservation status: Commercially threatened (IUCN Category).
Part(s) used: Gum of the bulb.

Uses: Gum of the bulb is used to apply on sores.

Cordyceps sinensis (Berk.) Sacc.

Common name(s): Yarsagumba (Nep.); Jeewan buti (Dolp.)
Family: Hypocreaceae

Description: A club-shaped fungus parasitic, later becoming saprophytic on insect larva after its death. The whole fungal body together with the host larva is less than 15 cm long, growing on moist alpine meadow (Malla *et al.* 1997).

Distribution: From east to west Nepal and from subalpine to alpine regions.

Conservation status: Banned for collection, use, sale, distribution, transportation and export (under the Forest Act of HMG 1993).

Part(s) used: Whole fungus body together with host larva.
Uses: Aphrodisiac, tonic, cardiactonic, expectorant etc.

Coriaria nepalensis Wall.

Common name(s): Bhujinsin (Nepbh.) **Family:** Coriariaceae
Chromosome number(s): 2n=40

Description: A glabrous herb. Leaves usually broadly rounded, cordate and abruptly acuminate, rarely ovate, oblong or elliptic, acute or acuminate. Flowers green, small, in clustered racemes. Fruit a berry, black, one-seeded.

Distribution: WCE; alt. 1200-2400 m
Himalayas (Kashmir to Bhutan), NE India, N. Myanmar, W. China.

Part(s) used: Leaves

Uses: Poisonous in large doses.

Costus speciosus (J. König) Sm.

Synonym(s): *Costus nipalensis* Roscoe; *C. speciosus* var. *angustifolius* Ker Gawl.; *C. speciosus* var. *nipalensis* (Roscoe) Baker
Common name(s): Kusth (Nep.); Kemuka (Sans.); Spiral ginger (Eng.) **Family:** Zingiberaceae
Chromosome number(s): 2n=18, 36

Description: An aromatic herb with fleshy branched underground rhizome. Stem 1-2 m stout, red, leafy. Leaves oblong, acute, sheathing at the base, 15-30 cm. Flowers c. 4 cm across, white, funnel-shaped, with contrasting large bright red ovate bracts of 1.5-4 cm long, numerous in a very dense cylindrical spike of 5-10 cm long. Calyx with 3 oval lobes; corolla tube shorter than calyx with unequal petals measuring 2.5-4 cm; lip white with an orange red center, rounded, 5-8 cm, forming a funnel with margins incurved and meeting. Capsule red, crowned with persistent calyx.

Distribution: WCE; alt. 400-700 m.
Himalayas (Nepal, Sikkim), south to Sri Lanka, Indo-China, Malaysia to New Guinea, Taiwan.

Part(s) used: Roots.

Important biochemical constituent(s): Tubers contain large amount of starch and rhizome contains diosgenin and ligogenin (Anonymous 1994).

Uses: The root is used in bronchitis, fever, dyspepsia, inflammations, anaemia, rheumatism, lumbago, pain in the marrow and hicough. It is also used as tonic. It has anthelmintic, depurative and aphrodisiac properties.

Cryptolepis buchananii Buch.-Ham.

Synonym(s): *Crateva religiosa* auct.
Common name(s): Silpigan (Nep.)
Family: Capparaceae

Description: A tree of about 15 m tall with lenticellate branches. Leaves long petioled; leaflets ovate-lanceolate, to 12 cm long and 4.9 cm broad, acuminate, entire, glaucescent beneath. Inflorescence terminal corymb. Flowers large, white or yellow; pedicel *c.* 4.5 cm long. Fruit round, 2-2.5 cm across, fleshy with numerous seeds embedded in yellow pulp.

Distribution: CE; alt. 100-1800 m.
Himalayas (Nepal, Sikkim), NE India, Myanmar, Indo-China, S. China.

Conservation status: Rare (IUCN Category).
Part(s) used: Bark and leaves.

Uses: Bark is demulcent, stomachic, laxative, diuretic, antipyretic, alterative and tonic. It is useful in calculus afflictions. Decoction of leaves is used in urinary infection. Root bark and fresh leaves has rubifacient properties.

Dr. Kamal K. Joshi and Prof. Sanu Devi Joshi

Cryptolepis buchananii Schult.

Synonym(s): *Cryptolepis reticulata* Wall.
Common name(s): Kalo sariwa (Nep.); Krishnasariva (Sans.)
Family: Asclepiadaceae

Chromosome number(s): 2n=22

Description: A glabrous climbing shrub with milky juice. Leaves 6-13 cm long and 2.4-5 cm broad, oblong or elliptic, entire. Follicles 7-8 cm long, straight, gradually narrowing from the middle, divaricate. Seeds 0.6 cm long, oblong-ovate, contracted from the tip, dark brown when dry, with pappus at the tip.

Distribution: WCE; alt. 250-1500 m.
Subtropical Himalayas, India, Sri Lanka, Myanmar, east to W. & S. China.

Uses: Cooling, aphrodisiac, alterative, and tonic. Cures vomiting, fever, biliousness, diseases of blood etc.

Curculigo orchioides Gaertn.

Common name(s): Musali (Nep.); Mushali, Talamuli (Sans.); Black muscale (Eng.)
Family: Hypoxidaceae
Chromosome number(s): 2n=18, *c.*50

Description: It is a stemless herb with tuberous rootstocks. Root is black. Leaves are fleshy, elongated, lorate or ensiform (Bhattacharjee 1998).

Distribution: CE; alt. 500 m.
Himalayas (Kashmir to Assam), India, Sri Lanka, east to Japan, Malaysia, Australia.

Part(s) used: Rhizome.

Uses: Used in piles, jaundice, asthma, diarrhea. It is considered demulcent, diuretic, aphrodisiac, alterative. Also used as tonic and poultice for itch and skin diseases. Application of rhizome powder into cuts stops bleeding and dries up the wound.

Curcuma angustifolia Roxb.

Synonym(s): *Curcuma longa* auct.
Common name(s): Haledo (Nep.); Haloo, Besha (Npbh.); Haridra (Sans.); Turmeric (Eng.)
Family: Zingiberaceae
Chromosome number(s): 2n=32, 62, 64

Description: A perennial aromatic herb with fleshy root attaining 60 cm in height. Stem short with tufted leaves. Flowers yellow.

Distribution: C; alt. 1500 m.
Himalayas (Uttar Pradesh to Nepal), N. India.

Part(s) used: Rhizome.

Important biochemical constituent(s): Rhizome contains essential oil containing sabinene, cineol, borneol, Zingiberine and sesquiterpenes. More important constituents of rhizome is curcuminoide known for anti-inflammatory action (Anonymous 1994).

Uses: Aromatic, stimulant tonic, carminative, blood purifier, antiperiodic and alterative. Externally applied to sprains, wounds and injuries. Used in chest and abdominal distention and mucous discharge. Relieves congestion, rheumatalgia, irregular menses, amenorrhea. Decoction of rhizomes is used in purulent conjunctivitis and the fresh juice is anthelmintic and antiparasitic. Also used in many skin diseases.

Curcuma aromatica Salisb.

Common name(s): Jangali haledo (Nep.); Ban haloo (Npbh.); Vana-haridra (Sans.); Wild turmeric (Eng.)
Family: Zingiberaceae
Chromosome number(s): 2n=42

Description: A perennial herb with tuberous rootstock, tubers yellow and aromatic inside. Leaves elliptic, 90-120 cm and up to 20 cm wide, leaf-stalk as long as blade which is finely hairy beneath. Flowers pinkish white with an orange lip, in a dense leafless spike crowned with enlarged colored bracts tipped with pink and borne on a short stem with papery bracts, usually appearing before the leaves. Lower bracts green, funnel-shaped and encircling several flowers, which open in succession. Calyx short, cylindric; corolla tube funnel-shaped, petals unequal, the upper concave and broader; staminodes oblong; lip rounded with a reflexed swollen apex.

Distribution: CE; alt. 700-1100 m. Himalayas (Nepal, Sikkim), India, Sri Lanka.

Part(s) used: Rhizome

Important biochemical constituent(s): Protein, fat, carbohydrates, fibre, minerals, vitamins A, B, C and niacin. The rhizome contains 5-6% volatile oil composed mainly (about 58%) of turmerones and tertiary alcohol (9%).

Uses: It is used as flavorant, colorant and cosmetic. It is used in the preparation of medicinal oils, oitments and poultice. Turmeric is stomachic, blood purifier, vermicide, antiseptic, carminative and tonic. It is also administered as antiperiodic alternative in case of diabetes and leprosy. It is taken with warm milk or inhaled from boiling water to get relieved from sore throat and common cold. It is one of the chief components of cosmetics like turmeric creams.

Cynodon dactylon (L.) Pers.

Synonym(s): *Panicum dactylon* L.
Common name(s): Dubo (Nep.); Situ ghayen (Npbh.); Durva (Sans.); Bermuda grass (Eng.)
Family: Gramineae
Chromosome number(s): 2n=36, 40
Description: A small herb with creeping rhizome. Leaves to 10 cm long and 0.2 cm broad, glaucous, acuminate with hairy rim forming ligule. Flower single in spikelet. Spikes 1.5-4 cm, digitate; glumes 1-nerved, ovate-acute, lemma and palea sub-equal, equaling spikelet.
Distribution: WCE; alt. 100-3000 m. Widely distributed in all warm countries.
Part(s) used: Whole plant.

Uses: Used in haemorrhage, inflammation (of limbs and urinary tracts) and gastric disorders.

Cyperus rotundus L.

Common name(s): Mothe (Nep.); Masta (Sans.); Nut grass (Eng.)
Family: Cyperaceae
Chromosome number(s): 2n=108

Description: A perennial, stoloniferous herb with erect stem, to 70 cm, triquetrous. Inflorescence simple or compound umbel. Bracts 3, rays 3-9, spreading each *c.* 10 cm long. Spikelets straw-colored to reddish-brown, linear-lanceolate, 10-50-flowered, compressed; glumes ovate, keeled. Nuts oblong-obovoid, black.

Distribution: WCE; alt. 300-2400 m. Cosmopolitan.

Part(s) used: Whole plant.

Important biochemical constituent(s): Essential oil, mono and sesquiterpines, fatty acids and alkaloids.

Uses: The plant is diuretic, emmenagogue, anthelmintic, diaphoretic, astringent and stimulant. It is also used in stomach disorder, irritation of bowels, leprosy, fever, blood diseases, biliousness and dysentery.

Dactylorhiza hatagirea (D. Don) Soó

Synonym(s): *Orchis hatagirea* D. Don
Common name(s): Panchaunle
Family: Orchidaceae

Description: A terrestrial herb (ground orchid) with finger-like tubers, to 90 cm tall. Leaves oblong-lanceolate, to 30 cm long. Flowers in dense cylindrical spike, *c.* 1.8 cm long including the stout, curved, cylindrical spur, sepals and petals nearly equal, lip rounded, shallowly 3-lobed and with dark purple spots.

Distribution: WCE; alt. 2800-3960 m.
Pakistan, Himalayas (Kashmir to Bhutan), China (Xizang).

Part(s) used: Root

Uses: Decoction is given in colic pain. Root powder is used to releave fever. Root is also used in urinary troubles. It is an astringent, expectorant, nervine tonic, aphrodisiac and demulcent. Also used as farinaceous food.

Daphne bholua Buch.-Ham. ex D. Don

Synonym(s): *Daphne cannabina* Lour. ex Wall.; *D. papyracea* Wall. ex Steud.
Common name(s): Kagatpate (Nep.); Paper Daphne (Eng.)
Family: Thymelaeaceae

Description: An evergreen shrub *c.* 2 m tall with a smooth gray bark. Leaves 6-7 cm long, lanceolate, slightly leathery and with short stalk, crowded towards the end of branches. Flowers white, sweet scented in stalkless bunches. Fruits *c.* 1.5 cm long, oval and succulent, black or purple when ripe.

Distribution: CE; alt. 2000-2900 m.
Himalayas (Nepal to Arunachal Pradesh), NE India (Meghalaya), W. China.

Part(s) used: Whole plant.

Uses: Used as purgative and febrifuge.

Datura metel L.

Synonym(s): *Datura fastuosa* L.
Common name(s): Kalo dhatura (Nep.); Hakugu Dudhale (Npbh.); Dhustura (Sans.); Thorn apple, Downy datura (Eng.)
Family: Solanaceae
Chromosome number(s): $2n=24-48$

Description: A perennial tropical herb with tomentose erect stem, to 1.2 m high, stout, herbaceous, terete. Leaves 15-20 cm long, ovate-lanceolate, acute, unequal at the base and often cordate, entire or repand-denate, densely tomentose on both surfaces and generally glandular; petioles 6-9 cm long; peduncles erect at first and nodding afterwards. Calyx *c.* 7.5 cm long, inflated towards the middle, persistent and reflexed in fruit; teeth lanceolate, acuminate; corolla unequal, about twice as long as the calyx, white, tinged with green below, pubescent outside, limb 10-toothed. Fruit septicidal capsule globose, covered with slender spines.

Distribution: WCE; alt. 300-1200 m.
Tropical America, widely cultivated and naturalized elsewhere.

Part(s) used: Whole plant

Important biochemical constituent(s): Hyoscine

Uses: The plant is antiseptic, narcotic, sedative and is useful in asthma. Poultice made from its leaves is used to reduce inflammation of breast caused by excessive lactation.

Datura stramonium L.

Common name(s): Dhatura (Nep.); Dudhale, Dhatura (Npbh.); Dhattura (Sans.); Thorn apple, Devil's apple, Jimson weed (Eng.)
Family: Solanaceae
Chromosome number(s): 24, 48

Description: A glabrous annual herb to *c*.1.2 m tall. Leaves to 18 cm long and 12 cm broad, ovate, base unequal, sinuate-toothed, acuminate. Flowers few, large, erect, white, rarely violet, narrow, funnel-shaped, solitary, to 12 cm long and 7.5 cm across, with 5 lobes and long-pointed apices; calyx tubular, 5-ribbed and lobed, about half as long as corolla. Capsule ovoid, 4-valved, 3.5-7 cm long, covered with sharp slender spines, and with the enlarged base of the calyx below.

Distribution: WCE; alt. 200-2200 m.
Tropical America, widely cultivated and naturalized.

Part(s) used: Roots, leaves, flowers and seeds.

Important biochemical constituent(s): Hyoscine and Hyocyamine

Uses: It is antispasmodic in asthma and Parkinson's disease. To treat asthma, dried leaves of **D. stramonium** is smoked in a pipe or homemade cigarrette. Hyoscyamine hydrobromide is given in delirium, tremor and mania as well as during surgery and childbirth as pre-anaesthetic medicine. It is also effective in ophthalmology and motion sickness.

Flowers are used to make poultice which is used to apply in wounds. Leaves are also applied in boils and ulcers. Decoction of roots and flowers makes sedative. The plant especially the seeds are dangerous hallucinogen.

Delphinium himalayai Munz.

Synonym(s): *Delphinium himalayense* N.P.Chowdhury **Common name(s):** Atis (Nep.)
Family: Ranunculaceae

Description: A herb 40-60 cm tall. Leaf-blades to 10 cm wide, 5-lobed nearly to base, the lobes broad and further lobed and toothed. Flowers purplish-blue 2-2.5 cm including spur to 15 mm in a long one-sided spike 10-15 cm; outer petals bristly-hairy outside, hairless inside; inner petals blackish; spur ascending; flower-stalks erect. Follicles densely hairy.

Distribution: WC; alt. 3000-4500 m. Nepal.

Part(s) used: Roots and rhizome.

Uses: Roots are used in medicines to cure cough, diarrhea and ailments related to blood (Kattel 2000). Rhizome is given in cough and fever. Also used in liver trouble. Slightly poisonous. Detoxified before used in medicine (Ghimire *et al.* 2000).

Dendrobium nobile Lindl.

Common name(s): Noble dendrobium
Family: Orchidaceae
Chromosome number(s): 2n=20, 40
Description: An erect perennial epiphytic orchid to 50 cm tall. Stem jointed, compressed, solid, yellowish rather deeply furrowed. Leaves sessile, oblong,

obliquely notched. Flowers 2-4, subracemous on a short peduncle from a leafing or leafless stem, purple or white with purple tips and lip.
Distribution: E.
Himalayas (Nepal to Bhutan) NE India, China, Myanmar, Thailand, Indo-China.
Part(s) used: Stem.
Uses: Stem is useful in thirst and dryness of the tongue. It is also given in weakening and fever during convalescence (Anonymous 1989).

Desmostachya bipinnata (L.) Stapf

Synonym(s): *Eragrostis cynosuroides* (Retz.) P. Beauv.
Common name(s): Kush (Nep.); Kush (Npbh.); Kush, Yagyabhooshan, Soochyagra (Sans.)
Family: Gramineae

Description: A perennial tufted creeping grass to 90 cm tall and with stout rootstock and stolon. Leaves linear to 50 cm long and 1 cm broad, rigid with filiform lip and hispid margins. Panicles erect, columnar, to 45 cm long and 4 cm across, rachis puberulous, spikelets sessile. Involucral glumes unequal, lower smaller than the upper ones. Floral glumes *c.* 0.2 cm long, ovate.

Distribution: C; alt. 100 m.
North and tropical Africa, Iran, Arabia, Nepal, India.

Part(s) used: Whole plant.

Uses: The plant is sweet, acrid, cooling, aphrodisiac, diuretic. It is given in asthma, jaundice, biliousness and diseases of blood.

Dichroa febrifuga Lour.

Common name(s): Antifebrile dichroa (Eng.) **Family:** Saxifragaceae
Chromosome number(s): 2n=36

Description: An erect shrub about 1 m tall. Leaves opposite, petiolate, oblong, 7-12 cm long and 2-4 cm broad, acuminate, with attenuated base, serrate, glabrous or slightly villous. Inflorescence a compact, axillary or terminal panicle. Flowers numerous, compact, usually blue, 8 mm wide. Fruit a blue berry, 5 mm in diameter. Seeds very numerous, pyriform, small (barely 1 mm long).

Distribution: CE; alt. 900-2400 m.
Himalayas (Nepal to Bhutan), India, Myanmar, east to C. China, Taiwan, Malaysia.

Part(s) used: Roots.

Important biochemical constituent(s): a, b, g-dichroines are the three effective constituents, g-dichroine being the most potent factor.

Uses: Used in the treatment of malaria and productive cough.

Dioscorea bulbifera L.

Synonym(s): *Dioscorea sativa* auct.; *D. versicolor* Buch.-Ham ex Wall. **Common name(s):** Tarul (Nep.); Tarul (Npbh.); Varahi (Sans.); air potato, potato yam, aerial yam, bulb-bearing yam (Eng.)
Family: Dioscoreaceae
Chromosome number(s): 2n=80

Description: A perennial twining herb with tuberous root and numerous aerial brown and warty bulbils of irregular shape and size. Stem twining to the left. Leaves alternate, ovate to cordate, 7-14 cm long and 6-13 cm wide, acuminate, entire 7-11 nerved, glabrous. Flowers unisexual, dioecious, male spikes 5-10 cm long, clustered, axillary or in leafless panicles; perianth-segments 6; stamens 6. Female spikes 10-15 cm long, in axillary clusters of 2-5. Capsule 1.8-2.2 cm long, oblong. Seeds winged at the base.

Distribution: WCE; alt. 150-2100 m.
Tropics of the Old World.
Part(s) used: Tubers
Uses: They are given in haemoptysis, epistaxis, pharyngitis, goitre, pyogenic infections, scrofula, orchitis, sprains and injuries (Anonymous 1989).

Dioscorea deltoidea Wall. ex Griseb.

Synonym(s): *Dioscorea nepalensis* Sweet ex Bernardi
Common name(s): Bhyakur (Nep.); Yam (Eng.)
Family: Dioscoreaceae
Chromosome number(s): 2n=20

Description: A climber with unarmed, slender, round stem. Leaves simple, subdeltoidly cordate, caudate-acuminate, 7-9-nerved, membranous; petiole very slender and long. Male flowers in long, axillary, nearly simple spikes, perianth subrotate.

Distribution: WCE; alt. 450-3100 m.
Himalayas (Kashmir to Assam), Indo-China, W. China.

Conservation status: Commercially threatened (IUCN Category).
Part(s) used: Tuberous rootstock.
Important biochemical constituent(s): Diosgenin.

Uses: Used as fish poison and to kill lice. It is also used to extract diosgenin for the manufacture of steroid hormones and cortico-steroids.

Dioscorea prazeri Prain & Burkill

Synonym(s): *Dioscorea sikkimensis* Prain & Burkill
Common name(s): Bhyakur (Nep.); Bahrahkand (Sans.); Deltoid yam (Eng.)
Family: Dioscoreaceae
Chromosome number(s): 2n=20

Description: A glabrous creeper with short and stout rhizome and ridged unarmed stem. Leaves cordate or long-cordate, acuminate or with short very acute acumen, dorsal side shining and smooth. Male flowers sessile, 1-3 in spike-like inflorescence; female flowers up to 20 in slender spikes of *c.* 30 cm long. Perianth members shorter and thicker than in the male flowers. Capsule-wings broadly half-obcordate-obovate or subrhomboidal.

Distribution: WC; alt. 910 m.
Himalayas (Nepal to Bhutan), NE India, Myanmar, Indo-China, N. Malaya.

Dr. Kamal K. Joshi and Prof. Sanu Devi Joshi

Conservation status: Commercially threatened (IUCN Category). **Part(s) used:** Tuberous rootstock.
Important biochemical constituent(s): Diosgenin.

Uses: Used as fish poison and to kill lice. It is also used to extract diosgenin for the manufacture of steroid hormones and cortico-steroids.

Diploknema butyracea (Roxb.) H. J. Lam

Synonym(s): *Aesandra butyracea* (Roxb.) Baehni; *Madhuka butyracea* (Roxb.) J. Macbr.
Common name(s): Chyuri (Nep.); Lhuchi (Npbh.); Indian butter tree (Eng.)
Family: Sapotaceae
Chromosome number(s): 2n=18, 22

Description: A tree with dark gray or brownish bark. Leaves to 24 cm long and 8 cm broad crowded near the tip of the branches, petiolate, obovate-oblong, subobtuse, rhomboid at the base, entire, tomentose when young, becoming glabrous or floccose-tomentose beneath when mature. Flowers long-stalked, tomentose, white, crowded below the subterminal leaves. Calyx coriaceous, rusty-villous, 5-lobed, ovate, imbricate. Stamens many, inserted at the mouth of the corolla tube; anthers linear-lanceolate. Fruit a fleshy berry. Seeds few or mostly solitary.

Distribution: CE; alt. 200-1500 m.
Subtropical Himalayas (Uttar Pradesh to Arunachal Pradesh).

Part(s) used: Seeds.

Uses: Fatty acid (oil) from seeds is used as ointment in rheumatism. Also used as emollient for chapped hands and feet in winter.

Dryopteris filix-mas (L.) Schott

Common name(s): Unyu (Nep.); Male fern (Eng.)
Family: Polypodiaceae

Chromosome number(s): n=82, 2n=164

Description: An erect terrestrial fern with short and stout rhizome. Lamina bipinnatifid to decompound, veins free, forked. Sori normally dorsal on the veins.

Distribution: WCE; alt. 1200-2700 m
Part(s) used: Rhizome
Uses: Used as anthelmintic

Eclipta prostrata (L.) L.

Synonym(s): *Eclipta alba (L.)* Hassk.; *E. erecta* L.; *E. prostrata (L.)* Roxb.
Common name(s): Bhrigaraj, Bhangerijhar (Nep.); Bhinlaye (Npbh); Yerba de tajo (Eng.)
Family: Compositae
Chromosome number(s): 2n=18, 22

Description: An erect or prostrate herb with stems branching strigose and appressed white hairs. Leaves 1-5 cm long and 0.3-1 cm broad, opposite, sessile, oblong-lanceolate, sub-entire, narrowed at both ends, densely strigose-pilose on both surfaces. Heads radiate, 0.5-1 cm in diameter, solitary or 2 together on axillary and terminal peduncles. Involucre bracts 8, ovate, obtuse or acute, strigose with appressed white hairs. Ray florets many, ligulate; ligules small, spreading, entire, white. Disc-florets tubular, corollas often 4-thoothed. Achenes trigonous in the ligule, compressed, tetragonous in disc florets, ribbed on the margin (Malla *et al.* 1986).

Distribution: WCE; alt. 200-1200 m. Pantropical.

Part(s) used: Whole plant.

Important biochemical constituent(s): Polyacetylenic thiophenes, traces of nicotine and phytosterol A (Rajbhandari *et al.* 1995).

Uses: Tonic, deobstruent in hepatic trouble and spleen enlargement. Plant juice is given to treat fever, urinary and spleen trouble, paste is applied on wound and skin diseases. Roots emetic, purgative, applied externally as antiseptic to ulcers and wounds in cattle.

Elaeocarpus sphaericus (Gaertn.) K. Schum.

Synonym(s): *Ganitrus sphaericus* Gaertn.; *Elaeocarpus ganitrus* Roxb. ex G. Don
Common name(s): Rudrakshya (Nep.); Rudraksha (Npbh.); Rudraksha (Sans.); Utrasum bead tree (Eng.)
Family: Elaeocarpaceae
Chromosome number(s):

Description: A lofty tree with simple, elliptic, serrulate, glabrous leaves. Flower white with petals laciniate half way down; stamens 35-40. Fruit a drupe, globose, purple when ripe.

Distribution: CE; alt. 700-1700 m. Nepal, India, Malaysia.

Conservation status: Vulnerable (IUCN Category). **Part(s) used:** Fruits.

Important biochemical constituent(s): The stones rudrakshyas are collected from many species of *Elaeocarpus.* Some of them contain elaeocarpidine (Anonymous 1994).

Uses: Used in diseases of the head and epileptic fits.

Embelia tsjeriam-cottam (Roem. & Schult.) A. DC.

Synonym(s): *Embelia robusta* Roxb.
Common name(s): Vayuvidanga (Nep.), Vayuvidanga (Sans.)
Family: Myrsinaceae

Description: A rambling shrub or small tree with glabrous branches but the very young ones sometimes rusty tomentose. Leaves to 12 cm long and 6 cm broad, elliptic, shortly acuminate, gland-dotted, lower surface reddish and hairy on nerves. Flowers pale greenish yellow and minute. Fruits small, globose, apiculate with the style, and red when ripe.

Distribution: WCE; alt. 400-1600 m.
Subtropical Himalayas (Kashmir to Sikkim), India to Indo-China.

Part(s) used: Fruits.

Uses: Dried fruits have antibacterial and antitubercular activities. It is laxative and used in getting rid of tapeworms (Bhattacharjee 1998). Fruit is anthelmintic. It is also used internally for piles. It is sometimes used as an antispasmodic and carminative, and also in tuberculous gland of the neck (Kirtikar *et al.* 1981).

Entada phaseoloides (L.) Merr.

Synonym(s): *Entada scandens* Benth.
Common name(s): Pangra (Nep.); Pangra (Npbh.)
Family: Leguminosae
Chromosome number(s): 2n=28

Description: A very large woody climber with twice-pinnate leaves and paired woody tendrils. Leaves with 2 pairs of pinnae each with 3-4 pairs of oblong to obovate shining leaflets 3-8 cm long. Inflorescence long slender axillary spike *c.* 25 cm long or terminal branched cluster. Flowers tiny (2-3 mm across), pale yellow; petals 5, equal; stamens longer. Pod enormous, to 120 cm long and 10 cm broad, with 10-30 rounded or square flattened joints; seeds flat, shining, brown.

Distribution: E; alt. 350-1600 m. Tropical and subtropical Asia

Part(s) used: Seeds.

Important biochemical constituent(s): Saponins.

Uses: Seeds are used in pains of the joints, in debility and in glandular swellings. Internally used as an emetic.

Ephedra girardiana Wall. ex Stapf.

Common name(s): Somlata (Nep.); Kagchalo, Sallejari (Dolp.); Soma (Sans.)
Family: Ephedraceae
Chromosome number(s): 2n=14, 18

Description: A rigid tufted shrub to 60 cm tall with numerous densely clustered erect slender smooth green jointed branches arising from a branched woody base. Branches bear scales at the joints. Male cones ovate, 6-8 mm, solitary or 2-3, with 4-8 flowers each having 5-8 anthers with fused filaments and rounded fused bracts. Female cones usually solitary. Fruits ovoid 7-10 mm in diameter with fleshy red succulent bracts enclosing 1 or 2 seeds.

Distribution: W; alt. 3700-5200 m
Afghanistan, Karakoram, Himalayas (Kashmir to Nepal).

Part(s) used: Dried young branches and twigs.

Important biochemical constituent(s): Ephedrine and pseudoephedrine.

Uses: The plant is astringent and has strong pine odor. Ephedrine is prescribed in ailments like rhinitis, asthma, hay fever and emphysema, epilepsy, noctural enuresis, myasthenia gravis and urticaria accompanying angioneurotic oedema. Ephedrine salt as nasal spray is given in congestion and swelling. Subcutaneous injection is prescribed in hypotension caused by anesthesia. Pseudoephedrine is effective in nasal congestion. Ephedrine and pseudoephedrine are used in expectorants and bronchodilators.

Euphorbia hirta L.

Synonym(s): *Chamaesyce hirta (Linn.) Mill.* and *Euphorbia piluilifera* L.
Common name(s): Rato lahare ghans (Nep.)
Family: Euphorbiaceae
Chromosome number(s): 2n=12, 18, 20

Description: A prostrate or ascending herb with hispid stem to 60 cm long. Leaves opposite, shortly stalked, elliptic-oblong, obovate, toothed or serrulate. Inflorescence a cyathium, sessile or peduncled, axillary and subterminal, pubescent, with small, globose gland having very narrow or obsolete limb.

Distribution: WCE; alt. 150-1500 m. Pantropical.

Part(s) used: Whole plant.

Uses: The plants are prescribed for the treatment of colic troubles, dysentery, cough, asthama, vomiting and worms.

Ficus racemosa L.

Synonym(s): *Ficus glomerata* Roxb.
Common name(s): Udumber, Dumri (nep.); Udambara, Hemdugdhak (Sans.); Country fig tree (Eng.)
Family: Moraceae
Chromosome number(s): 2n=22, 26

Description: A tree of about 15 m tall with milky latex and gray bark. Internal surface of the bark red. Leaves to 15 cm long and 5 cm broad, upper surface dark green and lower surface pale. Fruits in stalked clusters, 2.5-5 cm in diameter, usually globular, red when ripe

Distribution: WC; alt. 300 m.
Pakistan, Nepal, India, Sri Lanka, China (Yunnan), Indo-China, Malaysia, Australia.

Part(s) used: Bark, latex and fruits.

Uses: Decoction of bark is gargled in mouth ulcer and is used to wash wounds. The bark powder and its decoction are used as stomachic. Paste of bark is applied locally in swellings and painful parts. Fruits relieve inflammation.

Flemingia strobilifera (L.) W. T. Aiton

Synonym(s): *Flemingia bracteata* (Roxb.) Wight; *F. fruticulosa* Wall. ex Benth.; *F. strobilifera* var. *bracteata* (Roxb.) Baker; *F. strobilifera* var. *fruticulosa* (Wall. ex Benth.) Baker; *Moghania strobilifera* (Linn.) J. St.
Common name(s): Shalparni (Nep.)
Family: Leguminosae
Chromosome number(s): 2n=22

Description: An erect shrub to 3 m tall with simple broadly lanceolate leaves. Leaves 5-13 cm long, silky-haired beneath. Bracts 1.2-2.5 cm long, rounded, pale green, folded papery, enclosing flower clusters. Flowers white, 2 or more in each bract, 9-10 mm long; calyx hairy, shorter than petals. Pod 9-10 mm long, densely hairy, cancelled by bracts.

Distribution: WCE; alt. 300-2300 m.
Tropics to subtropics in Asia, America (Jamaica, Trinidad).

Part(s) used: Leaves.

Important biochemical constituent(s): Quercitrin, rutin and quercimeritrin.
Uses: Vermifuge for children.

Foeniculum vulgare Mill.

Common name(s): Madhurika (Sans.); Fennel (Eng.) **Family:** Umbelliferae
Chromosome number(s): 2n=22, 44

Description: An aromatic perennial hardy herb of about 60 cm tall. Leaves 3 or 4 times pinnate, the ultimate leaflets very numerous, filiform, very elongated. Umbels compound, large, long-pedunculate, nearly regular. Flowers yellow, not involucrate. Fruit ovoid, 6 mm long and 2 mm in diameter, greenish; glabrous mericarp compressed dorsally, semicylindrical with 5 prominent, nearly regular ribs. Seeds oblong, ellipsoidal, to 6 mm long, straight or curved, greenish yellow.

Distribution: W; alt. 2300 m.
Nepal. Wide spread around the world, often as a result of escape from cultivation.

Part(s) used: Whole plant.

Important biochemical constituent(s): Volatile oil obtained from distillation of seeds contains d-pinene, camphene, dL-phellandrene, dipentine, anethol (50-60%), fencone, methyl chavicol, aldehyde and anisic acid.

Uses: Root juice is febrifuge, purgative, sudorific, and carminative. Leaves and leaf stalks are diuretic. Decoction of leaves strengthens the sight. Fruits and seeds are carminative and also used for flavor. Seeds facilitate digestion and its essential oil is aromatic. It cures asthma and colic. It is also used in chest, spleen and kidney troubles. Funnel is used as tonic and stomachic.

Fritillaria cirrhosa D. Don

Common name(s): Kalchelaharo (Nep.); Koylikaswan, Kwakhachola (Npbh.); Kakoli (Dolp.); Fritillary (Eng.)
Family: Liliaceae
Chromosome number(s): 2n=24

Description: A deep-rooted, slender, bulbous herb to 1 m tall. Stem erect, green, unbranched, with linear leaves, often long pointed, the upper ones with coiled tips. Flowers variable in color; maroon, yellow, green or purple, usually chequered, solitary, drooping, petals 3.5-5 cm long, narrow, elliptic, blunt. Capsule cylindric, 3 cm long and 2 cm across.

Distribution: WCE; alt. 3000-4600 m.
Himalayas (Nepal to Bhutan), China (Xizang), N. Myanmar.

Part(s) used: Dried bulbs.

Important biochemical constituent(s): Alkaloids and essential oil.

Uses: Given in asthma, bronchitis and tuberculosis.

Fumaria indica (Hausskn.) Pugsley

Synonym(s): *Fumaria vaillantii* var. *indica* Hausskn. **Common name(s):** Dhukure (Nep.); Araka, Kalapanga (Sans.) **Family:** Papaveraceae
Chromosome number(s): 2n=32

Description: A glaucous, leafy annual, to 30 cm tall, with much-branched delicate stem. Leaves 2-3-times cut into narrow pointed segments *c.* 1 mm wide. Flowers in clusters, small, pale pinkish to whitish, each 5-6 mm long. Sepals minute; upper petal with a short somewhat down-curved sac-like spur; flower stalks erect, equaling or slightly shorter than the lanceolate bracts. Fruit globular, *c.* 2 mm in diameter.

Distribution: E; alt. 200 m.
N. Africa, Europe, W. Asia, Himalayas, India, W. Siberia.

Part(s) used: Whole plant.

Uses: Decoction of the plant is used in jaundice. Also used to get relief from headache and fever.

Gardenia jasminoides J. Ellis

Synonym(s): *Gardenia augusta* Merr.; *G. florida* L. **Common name(s):** Cape jasmine (Eng.)
Family: Rubiaceae
Chromosome number(s): 2n=22

Description: An evergreen shrub to 2 m tall. Leaves opposite or ternately whorled, oblong-elliptical, 7-14 cm long and 2.5 cm broad, coriaceous, glossy, stipules membranous, often connate. Flowers solitary, terminal or axillary, broad, white, very fragrant. Fruit ovoid or ellipsoidal, 2.5-4.5 cm long, or orange, 5-8-ribbed and crowned by subulate calyx-teeth. Seeds numerous, oblong, compressed, 5 mm long and 3 mm wide, reddish, adhering to the placenta yellow-orange in color.

Distribution: CE: alt. 1000-1500 m.
Native of China and Japan, cultivated in Himalayas and India.

Part(s) used: Fruit.

Uses: The fruit is useful in the treatment of icteric infectious hepatitis, cold, fever, insomnia, conjunctivitis, mouth ulcer, toothache, epistaxis, haematemesis, haematuria, and externally in sprains and pain from stagnated blood (Anonymous 1989).

Gaultheria fragrantissima Wall.

Synonym(s): *Gaultheria fragrans* D. Don
Common name(s): Dhasingare, Machino, Patpate, Kolomba (Nep.); Wintergreen (Eng.)
Family: Ericaceae
Chromosome number(s): 2n=44

Description: A stout shrub with somewhat trigonous, glabrous, branches. Leaves lanceolate or ovate, shortly acute. Flowers in racemes, with bracts and bracteoles on the upper part of the pedicel; corolla green, greenish yellow or whitish, surrounded by the deep blue enlarged calyx; anther cell with 2 terminal bristles.

Distribution: WCE; alt. 1200-2600 m.
Himalayas (Nepal to Arunachal Pradesh), NE India (Meghalaya), N. Myanmar.

Part(s) used: Fresh plants.

Uses: Oil extracted from the plant is used to treat rheumatism. The oil added to certain ointment is liniment and counteracts its irritating effects. It has stimulant and carminative properties and used to treat hookworm (Bhattacharjee 1998).

Geranium nepalense Sweet

Synonym(s): *Geranium quinquenerve* Buch.-Ham ex D. Don; *G. radicans* DC
Family: Geraniaceae
Chromosome number(s): 2n=26, 28

Description: A slender pubescent diffused herb to 45 cm high, with roots arising from the nodes. Leaves opposite, stipules subulate-lanceolate, brown, petiole 0.3-6 cm long, hairy, palmately 3-5-lobed, acute, hairy on both the surfaces. Peduncles slender, 2-5 cm long, hairy, 1-2-flowered. Flowers light pink or pale purple, 0.5-1.5 cm across; sepals 5, equaling the petal, slightly awned, hairy; petal 5-lobed; stamen 10. Fruit an elongated, 5-lobed capsule, to 2.5 cm long.

Distribution: WC; alt. 1500-4000 m.
Afghanistan, Himalayas, NE India, Myanmar, N. Indo-China, W. China.

Part(s) used: Whole plant.

Uses: The plant is diuretic and astringent and used to treat afflictions related to kidney.

Gloriosa superba L.

Synonym(s): *Gloriosa doniana* Schult. & Schult. f.; *G. simplex* D. Don **Common name(s):** Langali (Sans.); Malabar glory lily (Eng.)
Family: Liliaceae
Chromosome number(s): 2n=22, 88

Description: A tall herbaceous plant attaining to 3-6 m. Leaves sessile or short- petioled, oblong to lanceolate, membranous, many-nerved. Flowers large, reddish-yellow, with narrowly lanceolate perianth segments, wavy edges, borne solitary or in sub-corymbose inflorescence towards the end of branches (Shrestha and Joshi 1996).

Distribution: CE; alt. 400-2200 m. Tropical Africa, Asia.

Conservation status: Rare (IUCN Category). A plant of very rare occurrence and is found as scape plants in vicinity of villages.

Part(s) used: Tubers.

Uses: Tubers possess abortifacient, stimulant and anthelmintic properties. It is used to treat leprosy.

Gmelina arborea Roxb.

Synonym(s): *Gmelina rheedii* Hook.
Common name(s): Khamari (Nep.); Gambhari (Sans.); Coomb teak (Eng.)
Family: Verbenaceae
Chromosome number(s): 2n=36, 38

Description: A deciduous, medium-sized tree, to 25 m tall. Leaves simple, opposite, long-stalked, broadly ovate, 7-15 cm long or broad, entire, acute, base truncate or cordate. Flowers usually appearing before leaves, dull chestnut colored with yellow markings, tubular to 4 cm long, borne in bunches at the ends of the branches. Fruit oblong or oval, succulent drupes, *c.* 3 cm long, yellow, with a hard stone.

Distribution: WE; alt. 200-1100 m.
Himalayas (Nepal to Bhutan), India, Sri Lanka, Philippines.

Part(s) used: Root, bark, leaf and fruit.

Important biochemical constituent(s): Yellow viscid oil, resin, an alkaloid with traces of benzoic acid.

Uses: Root and bark are bitter, stomachic, laxative and galactagogue. It is one of the ingredients of Dasamula (an Ayurvedic preparation comprising a group of roots of 10 plants). To avoid abortion at the early stage of pregnancy powder of bark mixed with majitho *(Rubia manjith)* seeds and satavari *(Asparagus racemosus)* is given with milk. Leaves are used as demulcent and are rubbed on forehead for the relief from pain. Fruit decoction is given in fever and biliousness.

Hedera nepalensis K. Koch.

Synonym(s): *Hedera helix* auct.; *H. himalaica* Tobler **Common name(s):** Dudela (Nep.)
Family: Araliaceae
Chromosome number(s): 2n=48

Description: A large evergreen woody climber. Leaves alternate, simple, linear lanceolate to cordate-ovate. Flowers yellowish green, polygamous. Fruit a berry, globose, yellow or red; seeds 3-5, ovoid.

Distribution: WCE; alt. 2000-3200 m
Afghanistan, Himalayas (Kashmir to Bhutan), NE India, Myanmar, China.

Part(s) used: Leaves and berries.

Important biochemical constituent(s): Triterpenoid saponins (Evans 1989).
Uses: Used as stimulant, diaphoretic and cathartic.

Hedychium spicatum Smith

Synonym(s): *Hedychium album* Buch.-Ham. ex Wall.
Common name(s): Kapurakachari, Sati (Sans); Spiked ginger lily (Eng.)
Family: Zingiberaceae

Chromosome number(s): 2n=34

Description: A herb to 150 cm tall, with robust leafy stems and fragrant flowers. Leaves oblong, to 30 cm long and 12 cm broad. Bracts large, green, 1-flowered; calyx papery, 3-lobed, shorter than bracts. Corolla tube 5-6.5 cm long, and much longer than calyx, petals white, linear, spreading; lip white with two elliptic lobes with an orange base; filament of stamen red. Capsule globular, 3-valved, with an orange-red lining; seeds black with a red aril.

Distribution: C; alt. 1800-2800 m.
Himalayas (Himachal Pradesh to Arunachal Pradesh.

Part(s) used: Rootstock.

Important biochemical constituent(s): Diterpenes (Evans 1989). Starch, cellulose, mucilage, albumen, acid resin and an essential oil have been isolated from the rootstock (Anonymous 1994).

Uses: Rootstock is fragrant and used in 'abir', a perfume. Powder or paste of rootstock is given with warm water in cough and asthma.

Hemerocallis fulva (L.) L.

Synonym(s): *Hemerocallis disticha* auct.; *H. lilioasphodelus* var. *fulva* L.
Common name(s): Orange day-lily, Tawny-day-lily (Eng.)
Family: Liliaceae
Chromosome number(s): 2n=22, 33, ?36

Description: A perennial herb 30-90 cm high with short fleshy rootstock. Leaves linear, 60-100 cm long and 2.5-4 cm broad, erecto patent, acute, subglaucous underneath. Scape 60-80 cm high, corymb, pedicels short, bracts small, membranous. Flowers inodorous, 7-12 cm in diameter. Tube yellow-red, outer segments orange-yellow, oblong, acute, inner much larger and broader than the outer; margins undulate with reticulate nerves.

Distribution: WC; alt. 2400-3600 m.
S. Europe, India, China, Japan, cultivated throughout India and possibly escaped from cultivation in Nepal.

Part(s) used: Roots and rhizomes.

Uses: The roots and the rhizomes are useful in the treatment of jaundice, cystitis and difficulty in micturition. They are used externally in breast abscess.

Heracleum nepalense D. Don

Synonym(s): *Heracleum nepalense* var. *bivittata* C.B. Clarke
Common name(s): Nafo, Bhote jeera (Dolp.)
Family: Umbelliferae

Chromosome number(s): 2n=42, 44, 48.

Description: A shrub to 160 cm tall with sparingly hairy stem. Leaves usually nearly hairless on undersides except for bristly hairs on the veins beneath. Lower leaves pinnate with toothed, deeply lobed, or pinnately-cut large leaflets, the upper leaves ternate with leaflets often 3-lobed. Flowers white; bracts 3-5, linear, inconspicuous, or absent. Fruit obovate, to 8 mm long and 6 mm broad, commissure, with broad lateral wings, weak ribs and 4 vittate.

Distribution: WCE; alt. 1800-3700 m
Himalayas (Uttar Pradesh, Nepal, Bhutan), NE India.

Part(s) used: Seeds.

Uses: Seeds are flavorant and used in various recipes.

Hippophae tibetana Schltdl.

Synonym(s): *Hippophae rhamnoides* auct.; *H. rhamnoides* subsp. *tibetana* (Schltdl.) Servett.
Common name(s): Dale chuk (Nep.)
Family: Elaeagnaceae
Chromosome number(s): 2n=12, 20, 24

Description: A densely branched shrub or a tree with long stout terminal spines formed from the tips of old branches. Leaves small narrow-elliptic, 1.5-2 cm long and 2-4 mm broad, numerous, covered with rust-colored scales. Flowers stalkless, yellowish, c. 4 mm across, in clusters appearing on leafless stems. Fruit orange-red when ripe.

Distribution: CE; alt. 3800-4500 m. Himalayas (Punjab to Bhutan), N & W China

Part(s) used: Ripe fruit.

Uses: Dilutes blood. Also used as tonic, appetizer and herbal tea. It is given in cold and cough.

Holarrhena pubescens (Buch.-Ham.) Wall. ex G. Don

Synonym(s): *Holarrhena antidisentrica* Wall. ex A. DC.; *H. codaga* G. Don
Common name(s): Indrajau, Bankhirro (Nep.); Kutaja (Sans.); Conessi or Tellicherry bark (Eng.)
Family: Apocynaceae
Chromosome number(s): 2n=22

Description: A dwarf tree of 10 m tall with milky latex. Leaves thin, ovate, to 30 cm long, nerves conspicuous and leaf stalk small. Flowers in large terminal branches, white and fragrant. Fruits linear, slender, cylindric, to 45 cm long and 1 cm thick, dark gray with white specks. Seeds 1 cm long.

Distribution: WCE; alt. 100-1500 m.
Tropical Himalayas, India, Myanmar, Indo-China, Malaya.

Part(s) used: Bark and seeds.

Important biochemical constituent(s): It contains primarily conessine alkaloid and other alkaloids are concuressine, conessimine, isoconessimine, holonamine, kurchessine, kurcholessine, kurchicine and a photoplasmic poison, emetine (Bhattacharjee 1998).

Uses: Bark is used as tonic and febrifuge and has digestive properties. Seeds are considered effective in diarrhea, fevers, jaundice and stone in bladder. Conessine is prescribed in the treatment of amoebic dysentery and vaginitis. It also found effective in suppressing the growth of tubercular bacilli.

Hygrophila auriculata (Schumach.) Heine

Synonym(s): *Hygrophila spinosa* T. Anderson
Common name(s): Talmakhan (Nep.)
Family: Acanthaceae
Chromosome number(s): 2n=32

Description: An unbranched stout erect herb of about 150 cm tall. Stems 4-angled and hairy. Leaves whorled, six at a node and each possesses sharp yellow colored spines on its axil. Flowers purple blue, about 3 cm long, 2-lipped, lower lip folded, 8 in a whorl.

Distribution: WCE; alt. 200-700 m.
Tropical Himalayas, India, Myanmar, Indo-China, Malaya.

Part(s) used: Whole plant.

Uses: Roots and seeds are diuretic. Leaves are used in cough and urethral discharges. Seeds are useful in the treatment of veneral diseases. The drug from plant is used to treat diseases of urinogenital system, jaundice, dropsy and rheumatism (Bhattacharjee 1998).

Hyoscyamus niger var. agrestis (Kit.) Beck.

Synonym(s): *Hyoscyamus agrestis* Kit.; *H. niger* subsp. *agrestis* (Kit.) Hultén
Common name(s): Bazar bhang (Nep.); Langthang, Langthang tse (Am.); Dipya (Sans.); Henbane (Eng.)
Family: Solanaceae
Chromosome number(s): 2n=34, 36

Description: An erect, biennial or annual herb to 80 cm tall with robust stem and a disagreeable odor. Leaves spreading, oblong-ovate, 15-20 cm, coarsely sinuate-toothed. Cauline leaves smaller, sessile, oval-oblong, sinuate, pinnatifid, lobes irregular, triangular, lanceolate. Flowers in terminal scorpioid cymes, or axillary, nearly entirely sessile. Fruit a capsule, enclosed in the persistent and enlarged calyx. Seeds small, compressed, nearly ovoid, slightly reniform, 1 mm in diameter, brownish gray.

Distribution: WC; alt. 2000-3400 m.
Europe, N. Africa, SW & C. China, N. America.

Dr. Kamal K. Joshi and Prof. Sanu Devi Joshi

Part(s) used: Seeds.

Uses: Seeds are useful in the treatment of gastric or intestinal cramps, diarrhea, prolapse of the rectum, neuralgia, cough, hysteria, mania, skin inflammation and boils. It can also be used externally in pain due to dental caries.

Ichnocarpus frutescens (L.) R. Br.

Common name(s): Kalo shariva (Nep.); Sariva, Syamalata (Sans.); Black creeper (Eng.)
Family: Apocynaceae
Chromosome number(s): 2n=20

Description: A large much branched twining shrub. Leaves to 7.5 cm long and 4 cm broad, elliptic-oblong, acute or acuminate, main nerves 5-7 pairs with finely reticulate venation. Flowers greenish white, numerous in axillary and terminal cymes.
Distribution: WCE; alt. 150-900 m.
Tropical Himalayas, India, Myanmar, Indo-China, Malaya.

Part(s) used: Roots, stalks and leaves.

Uses: The root is sweetish, cooling, aphrodisiac, cures vomiting, fever, biliousness, and afflictions of blood. The root is also used as tonic. Decoction of stalks and leaves are used in fevers.

Impatiens balsamina L.

Synonym(s): *Impatiens coccinea* Sims
Common name(s): Tyuri (Nep.); Laincha (Npbh.); Balsam (Eng.)
Family: Balsaminaceae
Chromosome number(s): 2n=10, 12, 14

Description: An annual erect herb to 90 cm tall with fleshy, cylindrical, glabrous or pubescent stem having swollen nodes. Leaves alternate, 5-15 cm long and 1.5-2.5 cm broad, lanceolate, acuminate, serrate, base cuneate and decurrent. Flowers pink or white, showy. Pedicels 1-3, axillary, shorter than the leaves. Sepals minute, ovate. Standard small, orbicular, retuse, horned. Wings broad, lateral lobes rounded, terminal sessile, very large. Lip small, boat-shaped, mucronate, Spur short or long, incurved. Capsule tomentose. Seed reticulate.

Distribution: WCE; alt. 1200-1900 m. Native of SE Asia, widely cultivated.

Part(s) used: Seeds

Uses: Seeds are used to treat amenorrhea and dystocia.

Inula cappa (Buch.-Ham. Ex D. Don) DC.

Synonym(s): *Inula eriophora* DC.; *Conyza cappa* Buch.-Ham. ex D. Don
Common name(s): Raasna, Gai tihare (Nep.); Raasna, Surabhi (Sans.)
Family: Compositae
Chromosome number(s): 2n=20, 40

Description: A shrub with stout branches, silky, villous or wooly. Leaves almost sessile, oblong or oblong-lanceolate, acute, toothed. Heads numerous, involucral bracts linear and rigid. Achenes silky. Pappus-hairs dirty white, equaling the tubular corolla.

Distribution: WCE; alt. 150-2500 m.
Himalayas (Uttar Pradesh to Bhutan), NE India to China, Thailand, Java.

Part(s) used: Whole plant.

Uses: Decoction of the plant is given in rheumatic pain. Leaf paste is applied locally to relieve from rheumatic pain.

Inula racemosa Hook. f.

Synonym(s): *Inula royleana* auct.
Common name(s): Puskaramul (Nep.); Puskara mula (Dolp.); Pukarmula, Puskara (Sans.)

Dr. Kamal K. Joshi and Prof. Sanu Devi Joshi

Family: Compositae
Chromosome number(s): 2n=20

Description: A stout perennial herb to 1.75 m tall with large short-stalked flower heads. Stem rough, grooved. Lower leaves narrowed to a winged leaf-stalk, blade elliptic-lanceolate, to 45 cm long; the upper leaves lanceolate, half-clasping stem with large basal lobes, all leaves with rounded teeth and densely hairy beneath. Flower heads 4-8 cm across, borne in a spike-like cluster, or inflorescence branched with one or two flower heads at the ends of the branches. Ray-florets slender, to 2.5 cm long; outer involucral bracts broad-traingular, somewhat leafy, woolly-haired, the inner oblong blunt, the innermost linear.

Distribution: WCE; alt. 2500-3700 m.
Afganistan, Pakistan (Chitral), Himalayas (Kashmir to Nepal).

Part(s) used: Roots.

Important biochemical constituent(s): Roots contain essential oil, a bitter principle and benzoic acid (Anonymous 1994).

Uses: Root is used as tonic and stomachic. It is also used as expectorant and resolvent in duration.

Ipomoea nil (L.) Roth.

Synonym(s): *Ipomoea hederaceus* auct.; *Pharbitis nil (L.)* Choisy
Common name(s): Smaller morning glory (Eng.)
Family: Convolvulaceae
Chromosome number(s): 2n=30

Description: An annual twining hairy herb. Leaves cordate-ovate, 3-lobed, downy, to 6 cm long, acute. Peduncles axillary, hairy, 2-3-flowered. Flowers large, light bright blue; stigma subglobose, large, glandular, 3-lobed. Capsule trilocular, 2 seeds in each loculus; seeds 6 mm long and 4 mm broad, triangular, smooth, black or light brown.

Distribution: WC; alt. 760-2000 m.
Probably a native of the New World tropics, now widely cultivated and naturalized in other tropical and temperate areas.

Part(s) used: Seeds.

Uses: Used in oedema ascitis due to liver cirrhosis, constipation, abdominal pain due to parasitic infection.

Iris decora Wall.

Synonym(s): *Iris nepalensis* D. Don; *I. Sulcata* Wall. **Common name(s):** Padam puskar (Nep.)
Family: Iridaceae
Chromosome number(s): 2n=24

Description: A rhizomatous herb with slender stem to 30 cm long. Leaves linear, to 6 mm broad. Flowers solitary or few, pale lilac, in a branched sometimes long stalked cluster; falls to 2.5 cm broad, stalked with a central yellow ridge-like crest, standards narrower and smaller; corolla-tube *c.* 4 cm long. Spathes narrow, to 6 cm long.

Distribution: WCE; alt. 1800-4000 m.
Himalayas (Bashahr to Arunachal Pradesh), NE India (Meghalaya), W. China.

Part(s) used: Rhizome.

Important biochemical constituent(s): Rhizome contains irisolone.

Uses: Rhizome deobstruent, aperient, diuretic, useful in bilious obstruction; used externally as an application to small sores and pimples.

Juglans regia var. kamaonia C. DC.

Synonym(s): *Juglans kamaonia (C.DC.)* Dode
Common name(s): Okhar (Nep.); Khosin (Npbh.); Aksoda (Sans.); Walnut (Eng.)
Family: Juglandaceae
Chromosome number(s): 2n=32, 34, 36

Description: A large deciduous tree with gray vertically fissured bark. Leaves pinnate; leaflets 5-13 in number, elliptic to ovate, pointed, entire, leathery to 20 cm long. Male catkins pendulous, green to 12 cm long; female flowers very small in a short spike. Fruits large green drupes containing wrinkled nuts.

Distribution: WCE; alt. 1200-2100 m.
Himalayas (Kashmir to Bhutan), NE India (Meghalaya), China (Xizang).

Part(s) used: Bark of root and stem, leaves, green rind of unripe fruit and oil of nuts.

Uses: Young stem is used as toothbrush that relieves toothache. The bark of both root and stem is anthelmintic. Extract of the bark is purgative. Decoction of leaves is used as astringent for leucorrhoea and lymphatic afflictions. Leaf is good remedy for scrofula. Oil from the nut is also anthelmintic and used in the treatment of skin diseases.

Juniperus indica Bertol.

Synonym(s): *Juneperus pseudosabina* auct. and *Juniperus wallichiana* Hook. f. & Thomas ex Brandis
Common name(s): Dhupi (Nep.); Black juniper (Eng.)
Family: Cupressaceae
Chromosome number(s): 2n=44

Description: A large gregarious shrub, or small tree to 20 m tall, with a stout trunk. Leaves two kinds; those on lower branches awl-shaped, 3-6 mm long, spreading, those on terminal branches scale-like, c. 1.5 mm long, adpressed, overlapping in four ranks giving a smooth cord-like appearance to the branches. Fruit 1-seeded, at first brown then shining blue, to 13 mm long.

Distribution: WCE; alt. 3700-4100 m.
Karakoram, Himalayas (Kashmir to Nepal), W. China

Part(s) used: Wood and fruits.

Uses: Used to make incense stick. The plant is bitter, acrid, heating, appetizer, carminative, anthelmintic, alexipharmic, laxative. It is given to treat diarrhea, abdominal pains, diseases of spleen and abdomen, ascites, tumors, piles, bronchitis, indigestion, constipation, vaginal discharges. Fruit is stomachic, aphrodisiac, stypic. It has a bad taste and useful in asthma, hemicrania, chronic bronchitis, ailments of liver and spleen. It is also used in hydrocele and prolapse of the rectum. Oil from the fruit is emmenagogue, a abortifacient, tonic, anthelmintic, good for the treatment of earache, toothache, piles and is cooling to the brain.

Juniperus recurva Buch.-Ham. ex D. Don

Synonym(s): *Juniperus excelsa* auct.; *Juniperus macropoda* auct. **Common name(s):** Dhupi (Nep.); Drooping juniper (Eng.) **Family:** Cupressaceae
Chromosome number(s): 2n=22

Description: A low spreading shrub, or a tree to 10 m tall or occasionally taller with rather lax growth. Stems often brown, with ultimate branches tail-like and curving separately in various directions. Leaves awl-shaped, 6-8 mm long, in whorls of three, more or less adpressed to the branchlets and loosely overlapping. Fruit purplish-brown to black, shining when ripe, ovoid, 8-13 mm long, 1-seeded.

Distribution: CE; alt. 3300-4600 m.
Pakistan (Chitral), Himalayas (Kashmir to Bhutan), NE India, Myanmar, W. China.
Part(s) used: Wood and fruits.
Uses: Used to make incense stick. The plant is bitter, acrid, heating, appetizer, carminative, anthelmintic, alexipharmic, laxative. It is given to treat diarrhea, abdominal pains, diseases of spleen and abdomen, ascites, tumors, piles, bronchitis, indigestion, constipation, vaginal discharges. Fruit is stomachic, aphrodisiac, stypic. It has a bad taste and useful in asthma, hemicrania, chronic bronchitis, ailments of liver and spleen. It is also used in hydrocele and prolapse of the rectum. Oil from the fruit is emmenagogue, abortifacient, tonic, anthelmintic, good for the treatment of earache, toothache, piles and is cooling to the brain.

Juniperus sibirica Burgsd.

Synonym(s): *Juniperus communis* var. *saxatillis* Pall. **Common name(s):** Dhupi (Nep.); Juniper (Eng.)

Family: Cupressaceae
Chromosome number(s): 2n=22

Description: An erect shrub to 1.5 m tall and in higher elevations a prostrate shrub. Leaves dense, needle-like sharp-pointed, 6-13 mm long, with a broad bluish-white band above, in whorls of 3. Male cones ovoid, looking like leaf buds. Female cones blue-black when ripe, 6-8 mm; seeds usually 3.

Distribution: WC; alt. 2700-3200 m.
Nepal, arctic and alpine regions of the N. hemisphere.

Part(s) used: Wood, foliage and fruit.

Important biochemical constituent(s): Volatile oil, sugar, resin, juniperin, fixed oil, protein, wax, gum, pectin, organic acids and salts.
Uses: The wood yields cedar oil and is used to cure chronic disorders of the urinogenital tract. Poultice made up of needles is useful in the treatment of wounds. Berries are used in aphonia, laryngitis and pharyngitis, pulmonary catarrh and asthma.

Justica adhatoda L.

Common name(s): Asuro, Vasaka (Nep.); Aleha (Npbh.); Vasaca, Malabar Nut (Eng.)
Family: Acanthaceae
Chromosome number(s): 2n = 34

Description: A large strong-smelling gregarious shrub. Flowers white, 2-lipped with red spots and streaks within, borne in short compact terminal and axillary spikes with conspicuous ovate overlapping bracts. Corolla c. 3 cm long, the tube short, the upper lip narrow, notched, incurved, the lower much broader, deeply 3-lobed; calyx with 5 short-stalked elliptic-lanceolate entire leaves 13-25 cm. Capsule club-shaped 3 cm, hairy, 4-seeded (Polunin and Stainton 1986).

Distribution: WCE; alt. 500 - 1600 m.
Subtropical Himalayas, India, Indo-China, Malaya.
Common on open and sunny place also cultivated as hedge plant.

Part(s) used: Leaf

Important biochemical constituent(s): Vasicine, essential oil, crystalline acid and a white crystalline alkaloid, (Rajbhandari *et al.* 1995).

Uses: Antiseptic, insecticidal, expectorant, anthelmintic, anti-inflammatory and antispasmodic. Used in chest diseases, tuberculosis, chronic bronchitis, asthma, and rheumatism.

Leonurus japonicus Houtt.

Synonym(s): *Leonurus sibiricus* auct.
Common name(s): Wormwood-like motherwort (Eng.) **Family:** Labiatae
Chromosome number(s): 2n=16, 18, 20

Description: An annual or biennial herb, to 1 m tall with erect, simple or branched, tetragonal, pubescent stem. Leaves opposite, long-petioled, palmate-tripartite, 7 cm long and 4 cm broad. Inflorescence a compact, axillary verticil. Flowers 2-lipped, upper lip oboval, curved, concave, lower lip equally long, expanded-trilobate. Fruit a collection of triquetrous nutlets.

Distribution: CE; alt. 200-2000 m.
Himalayas (Kashmir to Nepal), India, China, Japan, Malaysia

Part(s) used: Whole plant.

Important biochemical constituent(s): Leonurine
Uses: The plant is useful in the treatment of menstrual disturbances, amennorhoea, nephritic oedema, oliguria, haematuria, pyogenic infection. It is also used externally in ulcerous skin disease.

Lepidium apetalum Willd.

Synonym(s): *Lepidium ruderale* auct.
Common name(s): Kyaga (Am.); Darya ken (Am.) Pepperweed, Peppergrass (Eng.)
Family: Crucuferae
Chromosome number(s): 2n=32

Dr. Kamal K. Joshi and Prof. Sanu Devi Joshi

Description: An annual or biennial herb, to 35 cm tall, with erect or diffused stem densely covered with small capitate glandular hairs. Basal leaves pinnatisect, 3-5 cm long and 1-1.5 cm broad; cauline leaves sessile, lower leaves narrowly elongated, elliptic, dentate, upper linear, nearly entire. Inflorescence racemes. Flowers small, pedicel slender. Pods oval-elliptic, 2-5 mm long, compressed, flat, slightly emarginate, winged above. Seeds very small, oboval, light reddish brown or yellowish brown.

Distribution: WC; 2600-4100 m.
C. Asia, Himalayas, E. Siberia, Mongolia, China.

Part(s) used: Seeds.

Uses: The seeds are used in the treatment of abundant expectoration with cough and dyspnea, distention in the chest and hypochondrium, oedema and oliguria.

Leucas cephalotes (Roth) Spreng.

Synonym(s): *Leucas capitata* Desf.
Common name(s): Dronapushpa (Nep.); Gojapu swan (Npbh.); Chhatraka (Sans.)
Family: Labiatae
Chromosome number(s): 2n=22, 28

Description: An annual hairy herb to 90 cm tall. Leaves short-petioled, ovate or ovate-lanceolate, to 10 cm long and 3.5 cm broad, subacute, crenate-serrate, with tapering base, hairy on both surfaces, petiole to 2 cm long. Flowers sessile in globose dense terminal whorls, to 5 cm across; bracts lanceolate, 1.3-1.6 cm long and 0.2-0.5 cm broad, acute, awned, ciliate; calyx tubular, 1.5 cm long, slightly curved; corolla 2 cm long. Nutlet obovoid-oblong, 3 mm long, brown.

Distribution: WE; alt. 150-2400 m.
Afghanistan, Punjab, Himalayas (Himachal Pradesh to Bhutan), India.

Part(s) used: Whole plant.

Important biochemical constituent(s): ß-sitosterol and glucoside.

Uses: The plant is a stimulant, laxative, anthelmintic and an insecticide. It is used to treat bronchitis, jaundice, inflammation, asthma, paralysis, and leucoma. Flowers are used to relieve from cold and cough.

Ligustrum nepalense Wall.

Synonym(s): *Ligustrum bracteolatum* D. Don; *L. indicum* (Lour.) Merr.; *L. nepalense* var. *grandiflorum* (Wall.) Mansf.; *L. nepalense* var. *vestitum* C. B. Clarke
Common name(s): Keri (Nep.)
Family: Oleaceae
Chromosome number(s): 2n=46

Description: A shrub to 3 m tall or a small tree of 10 m or more with twigs having many close lenticels. Leaves smaller 2-6 cm, dark green, paler beneath, elliptic, pointed, entire, short-stalked. Flower clusters smaller, mostly 3-7 cm long, creamy white 5 mm across, with a short tube and 4 blunt triangular lobes, in finely hairy inflorescence; calyx cup-shaped, half as long as corolla tube, not or scarcely lobed; stamens 2. Fruits glaucous blue-black, globular.

Distribution: WCE; alt. 1200-2700 m.

Himalayas (Uttar Pradesh to Sikkim), Myanmar, Indo-China. **Part(s) used:** Leaves.

Uses: Diuretic and applied in poultices to bruises.

Lilium nepalense D. Don

Synonym(s): *Lilium ochroleucum* Wall. ex Baker
Common name(s): Ban lasoon, Khiraule (Nep.); Gun labha (Npbh.); Nepalese lily, Tiger lily (Eng.)
Family: Liliaceae
Chromosome number(s): 2n=24

Description: An erect leafy herb, to 1 m tall with many lanceolate leaves measuring 4.5-11.5 cm long and 1-3 cm broad. Flowers predominantly yellow, c. 15 cm long, sweet-scented, solitary or few, usually with a large or small brown purple zone within, corolla-tube greenish on the outside; petals spreading, recurved above the middle; stamens protruding out with reddish anthers.

Distribution: WCE; alt. 2300-3400 m. Himalayas (Uttar Pradesh to Arunachal Pradesh).

Part(s) used: Bulb.

Uses: The bulbs are aromatic and used as flavoring dishes.

Lindera neesiana (Wall. ex Nees) Kurz

Synonym(s): *Benzoin neesianum* Wall. ex Nees
Common name(s): Siltimur (Nep.); Katbasi (Npbh.)
Family: Lauraceae

Description: A spicy glabrous, deciduous tree. Leaves membranous, ovate, acute or acuminate, cuneate or cordate at base, 3-nerved. Inflorescence umbel, 5-7-flowered, solitary or clustered; flowers on tomentose pedicel, minute, green. Fruit globose, seated on persistent perianth.

Distribution: CE; alt. 1800-2700 m.
Himalayas (Nepal to Bhutan), NE India, Myanmar.

Part(s) used: Root, bark, leaves and fruits.

Uses: The plant is aromatic, spicy and carminative. Powders of root and bark are sued to relieve pain. Leaves and fruits are used in the treatment of skin diseases.

Litsea cubeba (Lour.) Pers.

Synonym(s): *Litsea citrata* Blume; *L. dielsii* (Lév.) Lév.; *L. kingii* Hook.
Common name(s): Siltimur (Nep.)
Family: Lauraceae
Chromosome number(s): 2n=24

Description: A small to medium-sized, evergreen tree to 20 m tall, with smooth dark gray bark mottled with lighter patches. Leaves lanceolate, acuminate, *c.* 14 cm long and 4 cm wide, crowded together at the end of branches, with stalks of *c.* 2.5 cm. Flowers small, yellow, borne in dense masses at the end of the twigs. Fruits lemon-scented berry enclosed in persistent calyx, black when ripe.

Distribution: CE; alt. 1000-2700 m.
Himalayas (Nepal to Bhutan), NE India, Myanmar, Indo-China, China, Taiwan, Java.

Part(s) used: Roots, bark and fruits.

Uses: Powdered root and bark are used to relieve pain. Fruits are used in the preparation of various medicines (Adrian and Storrs 1990).

Lobelia chinensis Lour.

Synonym(s): *Lobelia radicans* Thunb. **Common name(s):** Chinese lobelia (Eng.) **Family:** Campanulaceae
Chromosome number(s): 2n=42

Description: A perennial procumbent, glabrous herb, to 15 cm tall, with milky latex. Leaves linear or lanceolate, 1.2 -2.5 cm long and 2.5-6 mm broad, remotely toothed or sub-entire, sessile, often bifarous. Flower solitary; pedicel axillary, as long as or longer than the leaves. Calyx-limb 5-partite; corolla pale-purple, oblique, 2-lipped, upper lip bipartite, lower 3-lobed, tubes glabrous or obscurely pubescent. Capsule 4-6 mm long; seeds numerous, minute, ellipsoidal, slightly compressed.

Distribution: C; 1300-1500 m.
Nepal, NE India (Meghalaya), Thailand, Indo-China, China, Korea, Japan, Malaysia.

Dr. Kamal K. Joshi and Prof. Sanu Devi Joshi

Part(s) used: Whole plant.

Uses: The whole plant has a pungent taste. It is used in snakebites, boils, ascites from cirrhosis and schistosomiasis and nephritic oedema. It is also given in purulent infections, enteritis and diarrhea.

Lobelia pyramidalis Wall.

Synonym(s): *Lobelia nicotianaefolia* auct.; *L. wallichiana* (C.Presl.) Hook. f. & Thomson
Common name(s): Eklebir (Nep.), Lobelia (Eng.)
Family: Campanulaceae
Chromosome number(s): 2n=28

Description: Widely branched glabrous herbs. Leaves linear-lanceolate, finely serrulate, glabrous above. Inflorescence a raceme. Flowers purple-rose to whitish. Fruit a subglobose capsule.

Distribution: WC; 1100-2300 m.
Himalayas (Uttar Pradesh to Arunachal Pradesh), NE India, N. Myanmar, Indo-China.

Part(s) used: Leaves and flowering tops.

Important biochemical constituent(s): Lobeline, an alkaloid.

Uses: It is expectorant and is administered in asthma and chronic bronchitis to relieve spasm. When taken in small doses it induces vomiting, causes nausea, copious sweating leading to general relaxation.

Lonicera japonica Thunb.

Common name(s): Japanese honeysuckle (Eng.)
Family: Caprifoliaceae
Chromosome number(s): 2n=18

Description: A twining shrub, to 9 m tall, with brownish stem and the young branches slender and hairy. Leaves ovate or oval-oblong, to 8 cm long and 3 cm wide, obtuse or acuminate, entire, base rounded or nearly cordate, pubescent beneath. Inflorescence a two-flowered cyme, in the axils of the terminal leaves. Flowers fragrant, bracts leaf-like, broad-ovoid to elliptic, bractlets sub-orbicular. Calyx with ovoid tube, teeth 5, often irregular. Corolla bilabiate, white at first, turning gradually into yellow, 3-4 cm long, the upper lip quadrified, lower lip entire, the tube as long as the limb. Fruit a shining black, fleshy berry.

Distribution: C; 1400 m.
China, Korea, Japan; cultivated in temperate regions including Nepal.

Part(s) used: Floral buds.

Important biochemical constituent(s): Luteolin and *i*-inositol have been isolated from the flowers.

Uses: It is useful in influenza, infection at upper respiratory tracts, tonsillitis and acute cojunctivitis. It is also used in pyodermas, wound infection and cervical erosion.

Lycium barbarum L.

Synonym(s): *Lycium halimifolium* Mill.
Common name(s): Barbary boxthorn, Common matrimony vine, Duke of Argyll's tea-tree (Eng.)
Family: Solanaceae
Chromosome number(s): 2n=24, 36

Description: A spinous shrub to 3 m tall, armed with sharp conical spines which sometimes elongates and behaves like stem bearing leaves and flowers. Leaves very variable, sometimes solitary, oblong-lanceolate, to 4.5 cm long and 6 mm broad. Flowers solitary or in fascicles of 2-5; pedicels filiform, 6-13 mm long. Ovary ovoid-oblong, glabrous, seated in a large membranous, cup-shaped disc. Fruit a berry, bright red in color, 1-2 cm long and 6 mm in diameter. Seeds 2.5 mm diameter, discoid or subreniform, embedded in a soft glutinous viscid pulp, very minutely pitted, orange-yellow.

Distribution: C; 1600 m.
China; widely cultivated for hedges and naturalized elsewhere.

Part(s) used: Root bark and fruit.

Important biochemical constituent(s): Fruits contain zeazanthin.

Uses: Root bark is used in low fever due to pulmonary tuberculosis and cough. Fruit is used in seminal emission, sores on the lower part of the body due to poor function of the kidney. Fruits are considered to possess aphrodisiac properties as well.

Lycopodium clavatum Linn.

Common name(s): Nagbeli (Nep.); Common club moss (Eng.)
Family: Lycopodiaceae
Chromosome number(s): 2n=68, 102, 136

Description: A terrestrial trailing pteridophyte, with creeping stem, rooting at regular intervals along its length. Branches forked, ascending, densely leafy. Leaves adpressed, small, linear, broadly incurved, filiform at the tip, spreading. Strobili erect, cylindric, 2-7 cm long, with ovate pointed bract. Sporangia in vertical position, green.

Distribution: WCE.
From subtropical to temperate zones.

Part(s) used: Spores.

Important biochemical constituent(s): Fatty acids - palmitic acid and linoleic acid.

Uses: Spore of *Lycopodium clavatum* also known as 'vegetable sulphur' is used as a local desiccative, which prevents chapping the children skin. It is also used to dust the bed of the patients lying on the bed for a prolonged period and developed sores on the body due to continuous rubbing against bed or bed sheet. It has anti-rheumatic, diuretic, and anti-epileptic properties. Its decoction is used as aphrodisiac, appetizer and to control severe diarrhea. It is also given in urinary and digestive complaints.

Macrotyloma uniflorum (Lam.) Verdc.

Synonym(s): *Dolichos biflorus* auct.
Common name(s): Kulattha (Sans.); Horsegram (Eng.) **Family:** Leguminosae
Chromosome number(s): 2n=20, 22, 24

Description: A slender, suberect, low-growing, succulent, pubescent, annual bushy herb about 50 cm in tall, profusely branching at the base, branches intertwining among themselves or with plants branches of companion crop. Leaves trifoliate; leaflets membranous, pilose, entire. Flowers in bunches in axillary racemes. Pods sickle-shaped, 1.2-5 cm long and 0.6 cm broad, tipped with a persistent style. Each pod bears 5-7 small, flattened, rhomboidal, brown to red, black or mottled seeds.

Distribution: CE; alt. 450-2800 m.
A native of India, widely cultivated in the tropics.

Part(s) used: Seeds.

Important biochemical constituent(s): Plant enzymes eeclin, strepogenins, B-sitosterol, bulbiformin, linoleic acid, polyphenols, free sugars, and isoflavons have been isolated from this plant (Bhattacharjee 1998).

Uses: Seeds are regarded as tonic, diuretic and astringent. Decoction of seeds is used to treat bowel haemorrhage, leucorrhoea and other menstrual derangement, colic pain, bladder and kidney stones and scrofula. Infusion of whole seed is administered in hypertension. Soup prepared from seeds is given in sub-acute enlargement of spleen, liver and piles.

Maesa chisia Buch.-Ham ex D. Don

Synonym(s): *Maesa dioica* A. DC. **Common name(s):** Bilouni (Nep.) **Family:** Myrsinaceae
Chromosome number(s): 2n=20

Description: A very large shrub. Leaves lanceolate-elliptic, acuminate, base cuneate, primary nerves *c.* 12, some of the leaves distinctly toothed. Inflorescence a compound raceme, axillary, 3.5-7.5 cm long.

Distribution: WCE; alt. 1200-2600 m.
Himalayas (Nepal to Bhutan), NE India, N. Myanmar.
Part(s) used: Root bark.
Uses: Insecticidal and also used in syphilis.

Maharanga emodi (Wall.) A. DC.

Common name(s): Maharangi (Nep.)
Family: Boraginaceae

Description: A hispid herb, c. 30 cm tall, with long, woody tap root. Leaves oblanceolate, acute, entire, hispid on both surfaces. Flowers pink with blue base in axillary and terminal cymes; corolla narrow at the mouth. Filaments dilated at the base.

Distribution: WCE; alt. 2200-4500 m.
Himalayas (Uttar Pradesh to Bhutan), China (Xizang).

Part(s) used: Whole plant.

Uses: Root is used to dye hair. The plant has cooling, laxative, anthelmintic, alexipharmic properties. It is good in diseases of eyes and piles.

Mahonia napaulensis DC.

Synonym(s): *Berberis nepalensis (DC.)* Spreng.
Common name(s): Jamane mandro (Nep.); Mahonia (Eng.)
Family: Berberidaceae
Chromosome number(s): 2n=28

Description: A tree to 5 m tall, subsimple, bearing leaves terminally. Leaflets 2-12 pairs, oblong-ovate, to 12.5 cm long and 6 cm broad, spinous-serrate, palmately 3-5-nerved. Inflorescence a raceme. Flowers yellow. Fruit an elliptic berry.

Distribution: WCE; alt. 200-2900 m. Himalayas (Nepal to Bhutan), NE India.

Part(s) used: Bark and fruits.
Important biochemical constituent(s): Alkaloids such as umbellatine and neprotine.
Uses: Bark is antidysenteric, antidiarrhoeic. Berries are diuretic and demulcent in dysentery.

Mallotus philippinensis (Lam.) Müll.

Common name(s): Sindure, Rohini (Nep.); Sansuh, Phalisi (Npbh.) Kampilla, Kampillaka (Sans.); Kamala tree (Eng.)
Family: Euphobiaceae
Chromosome number(s): 2n=22

Description: An evergreen dioecious tree to 8 m tall. Leaves variable in shape and size, oval, oblong-oval or lanceolate to 22 cm long and 8 cm broad, usually entire, 3-veined at the base, petiole to 5 cm long. Flowers unisexual, male pale yellow, in clusters of to 20 cm long spike. Fruit 3-lobed, clustered or erect stalk, covered with bright crimson powder.

Distribution: WE; alt. 150-1800 m.
Himalayas (Uttar Pradesh to Bhutan), India, Sri Lanka, Indo-China, China, Malaysia, Australia, Polynesia.

Part(s) used: Roots, bark, leaves and fruits and seeds.

Important biochemical constituent(s): Rottlerin and isoallorottlerin.

Uses: Paste prepared from root is applied on painful parts in articular rheumatism. Root and leaves are used to treat skin diseases. Decoction of bark is used in abdominal pain. Crude powder from the exterior of the fruit is anthelmintic and used against intestinal parasites like threadworm, hookworm, and roundworm. The powdery coat of fruit mixed with water is effective in expelling bile and relieving from colic pain. The glands present on fruits are purgative, cathartic and carminative. It is also used in the treatment of bronchitis, enlargement of spleen, jaundice and piles. Paste of seeds is prescribed to apply on wounds and cuts.

Malva vercillata L.

Synonym(s): *Malva alchemillaefolia* Wall.; *M. neilgherrensis* Wight
Common name(s): Majino (Dolp.); Cluster mallow , curly-leaved mallow (Eng.)
Family: Malvaceae
Chromosome number(s): 2n=*c*.76, 84

Description: An erect biennial herb to 90 cm high with reniform or rounded, palmately lobed leaves. Inflorescence axillary cyme consisting of 4-5 reddish flowers borne on a short peduncle in each cluster. Fruit a capsule splitting into many one-seeded mericarps. Seeds flattened, with a V-shaped slit on one side.

Distribution: WCE; alt. 2100-3000 m.
Europe, Egypt, Abyssinia, Himalayas (Kashmir to Bhutan), India, China, NE Asia; occasionally cultivated.

Part(s) used: Roots, stems, leaves and seeds.

Important biochemical constituent(s): Malvalic acid has been identified in the oils from leaves and seeds (Anonymous 1962).

Uses: Roots are useful to treat tiredness of the lower limbs and body, perspiration caused by weakness or of unknown origin, rectocele and chronic nephritis. Stems and leaves are used in icteric type of hepatitis. Seeds are effective in ailments like infection of urinary system, lithiasis, obstruction of milk secretion.

Mangifera indica L.

Common name(s): Aanp (Nep.); On (Npbh.); Amra, Chuta (Sans.); Mango (Eng.)
Family: Anacardiaceae
Chromosome number(s): 2n=40, 42

Description: A magnificent spreading evergreen tree to *c.* 25 m tall bearing a dense mass of glossy, dark green, narrowly elliptic or lanceolate, to 40 cm long and 10 cm broad, somewhat leathery, tapering leaves. Flowers small, pinkish white, in clusters borne in large panicles at the end of the branches, polygamous or mostly staminate, usually with a single stamen in each flower bearing pollens and the rest four reducing to staminodes. Fruit large, ovoid-oblong, asymmetrical drupe, to 30 cm long, with a thick or thin skin (exocarp), frequently dotted with prominent white lenticels. Mesocarp yellow, orange or red, edible. Endocarp thick, hard, fibrous, enclosing a single large flattened, exalbuminous, monoembryonic or polyembryonic seeds.

Distribution: WCE; 300-700 m.
Tropical Himalayas, India, Sri Lanka, Myanmar, Indo-China, Malaysia, widely cultivated and often naturalized in the tropics.

Part(s) used: Bark leaves, flowers, fruits and seeds.

Important biochemical constituent(s): Glucoside mangiferine in bark and leaves, essential oil with mangiferol and ketone along with many other chemical constituents in flowers, ß-carotene and xanthophyll in ripe mango and carbohydrate, protein, minerals, fats, amino acids etc. in seeds.

Uses: Apart from the nutritive values owing to the presence of vitamins A and C, soluble sugar and protein etc. in the ripe mango, it has various medicinal values in its bark, leaves, fruit, kernel and rind. The bark is astringent, has tonic action on the mucus membrane and used to treat diphtheria and rheumatism. Unripe mango is used in ophthalmia and eruptions. The rind is astringent, stimulating, and tonic in debility of stomach. Kernel juice is snuffed once a day for three days to stop nasal breeding.

Megacarpaea polyandra Benth.

Common name(s): Ruga sag (Dolp.) **Family:** Cruciferae

Description: A robust perennial herb of 1-2 m tall, radish-like odor. Roots thick and annulate. Flowers yellow and fragrant.

Distribution: WC; alt. 2700-4500 m. Himalayas (Kashmir to Nepal), China (Xizang).

Conservation status: Vulnerable (IUCN Category).

Part(s) used: Roots and leaves.

Uses: Roots are febrifuge and tonic and used in fever. Leaves are cooling and also used to cure malarial fever.

Melia azedarach L.

Common name(s): Bakena (Nep.); Khaibasi (Npbh.); Mahanimba (Sans.); China berry, Bead tree, Pride of India, Persian lilac, Indian lilac, Umbrella tree (Eng.);
Family: Meliaceae
Chromosome number(s): 2n=28

Description: A tree to 20 m tall with old purplish branches. Bark in irregular plates, tough-like or single quills about 6 mm in thickness; the outer surface marked with large lenticels and irregular longitudinal furrows, grayish-brown in color; the inner surface light yellow, the fracture plane fibrous. Leaves bipinnate, to 80 cm long; leaflets to 7 cm long and 3 cm broad, glabrous, ovate ot elliptic, acuminate, irregularly dentate. Inflorescence an axillary panicle, compound, divaricate, shorter than the leaves. Flowers fragrant, elongated, purple. Fruit a glabrous drupe with fleshy pericarp and ligneous endocarp. Seeds black, elliptical, tegument coriaceous.

Distribution: WE; alt. 700-1100 m.
Iran, Himalayas, east to China. Cultivated.

Part(s) used: Root bark or bark, heartwood, leaves, flowers and fruits.

Important biochemical constituent(s): $C_9H_{48}O$ comparable to santonin in bark. Bark also contains alkaloids azaridine and paraisine.

Uses: Root bark and bark are used in ascariasis, ancylostomiasis, ringworm. Leaf juice is considered anthelmintic, antilithic, diuretic and emmenagogue. Adecoction of leaf is regarded as astringent and stomachic. Aqueous extracts of the heartwood are useful in asthma. A gum collected from the tree is considered useful in spleen enlargement. A poultice of the flowers is applied to eruptive skin diseases and for killing lice. Fruits are poisonous, induce vomiting and develop symptoms of paralysis when eaten.

Mentha spicata L.

Synonym(s): *Mentha crispa* L.; *M. crispa* var. *exerta* Wall.; *M. pudina* Buch.-Ham. ex Benth.; *M. sativa* Roxb.; *M. viridis (L.)* L.
Common name(s): Pudina, Patana (Nep.); Nawa ghayen (Npbh.); Podinak, Pootiha, Rochani (Sans.); Spearmint (Eng.).

Family: Labiatae

Chromosome number(s): 2n=32, 36, 48

Description: An aromatic herb of *c.* 61 cm tall with runner underground stem. Leaves usually sessile or short petioled, ovate, to 3.8 cm long, acute, dentate. Flowers small, white, in spike.

Distribution: WC; alt. 1800-2700 m.
Europe, Afghanistan, Pakistan (Chitral), Himalayas (Kashmir to Nepal), China (Xizang), N. America, widely cultivated.

Part(s) used: Leaves and flowering tops.

Important biochemical constituent(s): Essential oil contains carvone, -pinen, -phellandrene, l-limonene, octyl alcohol and dihydrocarveol.

Uses: Leaves used, as tea is effective in neuralgia, indigestion, diarrhea and washing the sores. Its tea with cinnamon is prescribed for hastening childbirth at delivery. It is carminative and stimulant. It is also used as flavoring agent.

Mesua ferrea L.

Synonym(s): *Mesua nagassarium* (Burm. f.) Kosterm.; *M. roxburghii* Wight
Common name(s): Nagkesar (Nep.); Nagakeshara (Sans); Ironwood tree (Eng.)
Family: Guttiferae
Chromosome number(s): 2n=32

Description: A medium-sized to large evergreen tree with short trunk. Bark grayish or reddish brown, exfoliating in large thin flakes. Leaves lanceolate, coriaceous, generally covered with a waxy bloom underneath, red when young. Flowers large solitary or in clusters of 2-3, white, fragrant. Fruits ovoid, nearly woody, 2.5-5.0 cm long, with persistent calyx. Seeds 1-4, dark brown, up to 2.5 cm in diameter; cotyledons fleshy, oily.

Distribution: CE; alt. 400-900 m.
Nepal, India to Vietnam and Malay Peninsula, Sri Lanka, Andaman Island.

Part(s) used: Flowers and seeds.

Important biochemical constituent(s): Young fruits contain oleo-resin from which an essential oil is obtained. Seeds contain a fixed oil and two bitter principles (Anonymous 1994).

Uses: The flowers are astringent and stomachic. Dried flowers are given in vomiting, dysentery, cough, thirst, irritation of stomach, excessive perspiration and bleeding pile. Dried flowers are also used for the preparation of perfumed ointments. Seed oil is used in rheumatism and skin diseases.

Michelia champaca L.

Synonym(s): *Michelia aurantiaca* Wall.; *M.doltsopa* auct.
Common name(s): Champ (Nep.); Chaswan (Npbh.); Golden champa, Yellow champa, Magnolia (Eng.)
Family: Magnoliaceae
Chromosome number(s): 2n=38

Description: An evergreen tree of about 20 m tall with hairy branchlets. Bark smooth, pale gray. Leaves ovate to lanceolate, tapering to a long point, shining on upper surface, glabrous, or nearly hairy beneath. Flowers yellow, very fragrant, borne on short stalk. Carpels slightly stalked with hairy ovaries.

Distribution: CE; alt. 600-1300 m.
Nepal, India, Myanmar, Thailand, Indo-China, China (Yunnan); commonly planted in SE Asia.

Conservation status: Endangered (IUCN Category).
Part(s) used: Bark, leaves, flowers, fruits and seeds.

Uses: The bark is diuretic, febrifuge and stimulant. Leaf juice is used to treat colic. Oil extracted from the flowers is employed in perfumes. It is also used in cephalagia, ophthalmia, gout and rheumatism. Flowers and fruits are stimulant, diuretic, stomachic and antispasmodic. Seeds and fruits are used for healing of cracks in feet (Bhattacharjee 1998).

Mimosa pudica L.

Common name(s): Lajjawati, Lajwanti (Nep.); Lajja (Sans.); Sensitive plant (Eng.)
Family: Leguminosae
Chromosome number(s): 2n=48, 52

Description: A prickly spreading hairy herb with leaves very sensitive to touch. Leaves pinnate; pinnae usually 4 together arranged palmately. Flowers purplish-pink, in globular heads of 1 cm diameter, borne on slender stalks usually in axillary pairs. Fruit a pod, 1.5-2.5 cm long with 3-5 joints and covered with stiff bristles.

Distribution: CE; alt. 200-1200 m. Pantropical.

Part(s) used: Roots and leaves.

Important biochemical constituent(s): A toxic alkaloid mimosine ($C_8H_{10}O_4N_2$), tannin in the root and an adrenaline-like substance in the leaves.

Uses: The root is used in the treatment of asthma, fever, cough, dysentery and vaginal and uterine ailments. It is also used as antidote for snakebite. The root paste is used as poultice in abscess. Paste of leaves and roots is given in piles and diseases of kidney.

Momordica charantia L.

Synonym(s): *Momordica muricata* DC.
Common name(s): Tite karela (Nep.); Khai kakacha (Npbh.); Karavalli, Sushavi (Sans.); Bitter Gourd (Eng.)
Family: Cucurbitaceae
Chromosome number(s): 2n=22, 33

Description: An annual, slender climber. Stem 5-angled and furrowed. Leaves palmately 5-9 lobed. Fruit pendulous, 5-25 cm long, fusiform, ribbed with numerous tubercles. Seeds numerous tubercles, 1-1.5 cm long, brownish with scarlet aril.

Distribution: CE; alt. 300-2100 m.
Africa, tropical Asia. Widely cultivated.

Part(s) used: Roots, leaves and fruits.

Important biochemical constituent(s): An alkaloid - momordicine, a glucoside and a resin, an aromatic volatile oil and mucilage.

Uses: The fruit has long been used locally as a folk remedy for diabetes mellitus. It is also considered tonic, stomachic, carminative and cooling. They are prescribed in the treatment of rheumatism, gout and diseases of liver and spleen. It is also used in Ayurvedic medicine to cure ailments of respiration, dysentery (in children), and pyorrhea. The paste of fruits are applied in swellings.

Morchella esculenta (Linn.) Pers.

Common name(s): Guchhi chyau (Nep.); Morel mushroom, Sponge mushroom (Eng.)
Class: Ascomycetes
Family: Helvellaceae

Description: An ascomycetous fungus with white, thick, erect, tapering stalk bearing rounded, or conical pileus, yellowish brown to olive, covered with a number of prominent ridges. Appears on decaying vegetable matter in hill forests after melting of snow.

Distribution: WCE; alt.
Himalayas (Kashmir to Uttar Pradesh), Nepal.

Part(s) used: Whole plant.

Uses: Source of vitamins and considered a very nutritious mushroom.

Moringa oleifers Lam.

Synonym(s): *Moringa pterygosperma* Gaertn.
Common name(s): Sigru, Shobhanjana (Sans.); Drumstick tree (Eng.)
Family: Moringaceae
Chromosome number(s): 2n=14, 28

Description: A tree *c.* 10 m in tall with thick, soft, corky, deeply fissured bark, tomentose when young. Leaves usually tri-pinnate; leaflets elliptic. Flowers white, fragrant in large panicles. Pods pendulous, greenish, 22.5-50.0 cm or more in length, triangular, ribbed; trigonous with wings.

Distribution: CE; alt. 150-1100 m.
NW India, cultivated through the tropics. **Part(s) used:** Roots, leaves, flowers and seeds.

Important biochemical constituent(s): Spirochin (an alkaloid), pterygospermin (an active antibacterial principle) in roots, moringine, moringinine in root bark, antibacterial substance in leaves, traces of alkaloids, wax, quercetin, kaempferol etc in flowers and protein, minerals and fatty acidsin seeds (Anonymous 1962).

Uses: Roots of the young tree and bark of the root are vesicant and rubifacient. Roots are also used as condiments. Leaves are used to treat scurvy and catarrhal affliction. Paste of the leaves is applied externally on wounds. Flowers are tonic, diuretic and cholagogue. Seeds are antipyretic. Oil extracted from the seeds is used to treat gout and rheumatism.

Murraya koenigii (L.) Spreng.

Synonym(s): *Bergera koenigii L. and Chalacas koenigii (L.) Kurz.* **Common name(s):** Surabhininiba (Sans.); Curry leaf tree (Eng.) **Family:** Rutaceae
Chromosome number(s): 2n=18

Description: A small tree of to 5 m tall with dark gray bark. Leaves imparipinnate to 30 cm long; leaflets 11-25, obliquely ovate or rhomboid. Flowers white 1 in much branched terminal peduncled corymbose cymes. Fruit ovoid or subglobose, 6-10 mm diam. apiculate, rough with glands, black, seeded.

Distribution: WE; alt. 150-1450 m.
Himalayas (Uttar Pradesh to Sikkim), India, Sri Lanka, Myanmar, Indo-China, China. Frequently cultivated.

Part(s) used: Roots, stem, stem bark and leaves.

Important biochemical constituent(s): In addition to its nutritive importance owing to the presence of protein, carbohydrate, fibre, minerals, carotene, nicotinic acid vitamin C and calcium in its leaves, it has medicinal and aromatic values. Its leaves also contain crystalline glycoside, koenigin and a resin. Fresh leaves on stem distillation yield 2.5% volatile oil having aromatic and spicy note.

Uses: Roots, bark and leaves are carminative, tonic and stomachic. Juice of root is administered to relieve from pain associated with kidney. Bark is used externally to treat eruptions and bites of poisonous insects and reptiles. Leaves yield 0.04% of essential oil. Paste made from leaves is applied on utricaria (Bhattacharjee 1998).

Myrica esculenta Buch.-Ham. ex D. Don

Synonym(s): *Myrica farquhariana* Wall.; *M. integrifolia* Roxb. *M. nagi* auct.; *M. sapida* Wall.
Common name(s): Kaphal (Nep.); Käbasi (Npbh.); Box myrtle, Bayberry, Wax myrtle (Eng.)
Family: Myricaceae
Chromosome number(s): 2n=16

Description: A medium-sized tree with nodules in the roots capable of fixing atmospheric nitrogen due to the presence of *Frankia*. Leaves short-petioled, oblanceolate, acute or obtuse, entire, minutely gland-dotted beneath. Male flowers in drooping spikes; female flowers in axillary, erect spikes. Fruits globose, red succulent.

Distribution: WCE; alt. 1200-2300 m.
Himalaya(Kashmir to Bhutan), India, Myanmar, east to S. & W. China and south to Malaysia.

Part(s) used: Bark.

Important biochemical constituent(s): Myricetin (coloring matter in a form of glycoside), aglycone (a second glycoside in traces) in bark.

Uses: Astringent, carminative, and antiseptic. A decoction of the bark is used in asthma, diarrhea, fevers, affliction in lungs, chronic bronchitis, dysentery and diuresis. The bark is chewed to relieve from toothache and a lotion prepared from it is used to wash putrid sores (Anonymous 1962).

Nardostachys grandiflora DC.

Synonym(s): *Nardostachys gracilis* Kitam.; *N. jatamansi* DC.
Common name(s): Jatamansi, Balchhar (Nep.); Bhutle (Dolp.); Naswan (Npbh.), Spike nard (Eng.)
Family: Valerianaceae
Chromosome number(s): 2n=16, 32

Description: A perennial herb of about 75 cm tall with long woody rhizomes having aromatic fibers making rhizomes very fragrant. Lower leaves larger than the upper leaves. Flowers small pinkish white and arranged in bunches.

Distribution: WCE; alt. 3200-5000 m. Himalayas (Uttar Pradesh to Bhutan), W. China.

Conservation status: Vulnerable (IUCN Category). **Part(s) used:** Roots and rhizomes.

Uses: It is antispasmodic, stimulant and a tonic. It is prescribed for treatment of convulsion, certain types of fits, palpitation of heart and constipation. It is given to improve urination, menstruation and digestion. It is also used to color hair oil (Bhattacharjee 1998).

Neopicrorhiza scrophularifolia (Wall. ex Benth.) Hemsl.

Synonym(s): *Picrorhiza kurroa* auct. and *Picrorhiza scrophulariflora* Pennell
Common name(s): Kutki (Nep.); Picrorhiza (Eng.)
Family: Scrophulariaceae

Description: A creeper with stout rootstock covered with old leaf-bases above. Leaves 2-6 cm, oblanceolate and narrowed below to a winged leaf-stalk. Flowers dark blue to purple, in a dense cylindrical head, borne in a stout stem arising from a rosette of conspicuously toothed; calyx hairy, nearly as long as the corolla tube, calyx lobes 5, lanceolate, blunt; corolla *c.* 1.5 cm, with a long 3-lobed upper lip and a short lower lip; stamens and style exserted. Capsule 6-10 mm.

Distribution: WCE; alt. 3500-4800 m.
Himalayas (Uttar Pradesh to Bhutan), N. Myanmar, China (Sichuan, Xizang, Yunnan).

Dr. Kamal K. Joshi and Prof. Sanu Devi Joshi

Conservation status: Vulnerable (IUCN Category).
Part(s) used: Root, rhizome and stem.

Important biochemical constituent(s): Kutkin (a glucosidal principle), kurrin (a non-bitter product), kutkiol (an alcohol) and kutkisterol (Rajbhandari *et al.* 1995).

Uses: Rhizome is used to control fever and gastralgia. Root powder is used as laxative and also administered for the relief from abdominal pain. It is also prescribed in liver complaints, anaemia and jaundice.

Nyctanthus arbor-tristis L.

Common name(s): Parijata (Sans.); Palija swan (Npbh.); Sephalika (Sans.); Night jasmine (Eng.)
Family: Verbenaceae
Chromosome number(s): 2n=44, 46

Description: A tall shrub or a small tree to 9 m tall with drooping branches, and coarse and quadrangular stem. Leaves opposite, ovate to 15 cm long and 8 cm broad, acute, entire or with distant teeth, rough owing to the hairs emerging from bulbous bases on the upper surface. Flower white, sweet-scented, *c.* 2.5 cm across with slender orange corolla tube, borne in clusters of 3-5 at leaf axils at the terminal end of the branches. Fruits 2-seeded, compressed and orbicular capsules.

Distribution: WCE; alt. 200-1200 m.
Subtropical Himalayas, India; often cultivated for its fragrant flowers.

Part(s) used: Leaves and flowers.

Important biochemical constituent(s): Essential oil and nyctanthin.
Uses: Anthelmintic and laxative.

Nymphaea stellata Willd.

Common name(s): Nilkamal (Nep.), Blue water lily (Eng.) **Family:** Nymphaeceae
Chromosome number(s): 2n=28, 56, 84

Description: A perennial aquatic herb with a short ovoid, acute rootstock. Leaves peltate, 10-12 cm in diameter, orbicular or elliptic, entire, or obtusely sinuate-dentate, glabrous, often blotched with purple beneath. Flowers solitary, blue, white, light yellow or pink or purple. Fruit a spongy berry; seeds minute, longitudinally striate.

Distribution: E; alt. 200 m.
Africa, tropical Himalayas, India, Malaysia.

Part(s) used: Whole plant.

Uses: The powdered rhizome is used in dyspepsia, diarrhea and piles. An infusion of rhizomes and stems is an emollient and a diuretic and used in blennorrhagia and disease of the urinary tract. Macerated leaves are applied as a lotion in eruptive fevers. A decoction of flowers has narcotic effect. Seeds are stomachic and restorative (Anonymous 1962).

Ocimum tenuiflorum L.

Synonym(s): *Ocimum sanctum* L.
Common name(s): Tulasi (Nep.); Tulsi (Npbh.); Ajaka, Tulasi (Sans.); Holy or sacred basil (Eng.)
Family: Labiatae
Chromosome number(s): 2n=32, 32+0-3B, 34, 36

Description: An erect biennial or triennial herb of about 75 cm in height, profusely branched and hairy. Stems usually quadrangular. Leaves opposite, decussate, to 5 cm long, hairy, entire or toothed, dotted with minute glands. Bracts sessile; flowers small, less conspicuous, purplish or reddish, on slender spikes in small compact clusters. Fruits small; seeds reddish or yellowish, sub-globose.

Distribution: CE; alt. 400-900 m.
SW Asia, Himalayas (Nepal), Bangladesh, India, Sri Lanka, Myanmar, China, Thailand, Malaysia. Often cultivated.

Dr. Kamal K. Joshi and Prof. Sanu Devi Joshi

Part(s) used: Roots, leaves, tender shoots and seeds.

Important biochemical constituent(s): Essential oil of leaves and shoots contains eugenol as the major component. Other constituents are nerol, terpinene, pinene and cavacrol. Leaves also contain ursalic acid, apigenin, luteolin and orientin (Anonymous 1994).

Uses: Decoction of roots is given in malarial fever. Decoction of leaves is used to treat common cold. Juice of leaves is used in earache, digestive complaints, bronchitis, catarrh, skin diseases and to control ringworm. Oil extracted from the leaves has antibacterial and insecticidal properties (Bhattacharjee 1998).

Operculina turpethum (L.) Silva Manso

Common name(s): Nisoth (Nep.); Tribrit (Sans.); Turpeth root (Eng.)
Family: Convolvulaceae
Chromosome number(s): 2n=30

Description: A large climbing herb having milky juice. Roots long, branched, and fleshy. Leaves ovate, cordate, to 10 cm long and 7 cm broad. Flowers to 5 cm long, funnel-shaped, in few-flowered bunches, white. Fruits enclosed by sepals.

Distribution: W; alt. 600 m.
Widespread in Old World tropics from E. Africa to N. Australia and Polynesia, introduced in West Indies.

Part(s) used: Root with intact bark.

Important biochemical constituent(s): Turpethin
Uses: Powdered roots with bark are used as purgative.

Oroxylum indicum (L.) Kurz

Common name(s): Tatelo (Nep.); Shyonaka (Sans.) **Family:** Bignoniaceae
Chromosome number(s): 2n=28, 30, 38

Description: A tree to 12 m tall with thick barks. Leaves 60-120 cm across, ternately 2-pinnate, pinnules 3-5-foliate; leaflets ovate, 12.5 cm long, acuminate. Flowers pinkish brown, borne in a large terminal raceme on 30 cm long stalk. Capsules 30-90 cm long, flattened, curved, woody (Shrestha and Joshi 1996)

Distribution: WCE; alt. 400-1400 m.
Tropical Himalayas, India to Indo-China, Malaysia, W. & S. China.

Conservation status: Vulnerable (IUCN Category). **Part(s) used:** Root barks and fruits.

Important biochemical constituent(s): Bark contains oroxylin A, baicalein, scutellarein and scutellarein rutinoside. Seeds contain alkaloid, terpenes and saponins (Anonymous 1994).

Uses: The root bark is tonic and astringent and useful in diarrhea and dysentery. It is diaphoretic and is used in rheumatism. Boiled in sesamum oil, it has been recommended for otorrhoea. Tender fruits are refreshing and stomachic and the seeds purgative (Anonymous 1962).

Osbeckia nepalensis Hook.

Synonym(s): *Osbeckis chulesis* D. Don; *O. nepalensis* var. *albiflora* Lindl.; *O. speciosa* D. Don
Common name(s): Seto chulsi (Nep.)
Family: Melastomataceae
Chromosome number(s): 2n=38, 40

Description: A straggling branched shrub often hanging from rocks and banks, to 1 m or more with 4-angled stem and lanceolate entire leaves having 3-5 conspicuous parallel veins. Flowers mauve-purple or white, small 3-5 cm across, in dense branched clusters; calyx lobes lanceolate, hairless, calyx-tubes covered with large flat scales fringed with bristles. Fruit densely hairy, 1.2-1.8 cm, opening by pores at apex.

Distribution: CE; alt. 450-2300 m
Himalayas (Nepal to Arunachal Pradesh), NE India, N. Myanmar, Thailand, Indo-China, Malay, W. China.

Part(s) used: Flowers.

Uses: Paste of flowers is applied to sores in mouth of children.

Dr. Kamal K. Joshi and Prof. Sanu Devi Joshi

Osyris wightiana Wall. ex Wight

Synonym(s): *Osyris arborea* Wall.; *O. nepalensis* Griff.
Common name(s): Nun dhiki (Nep.)
Family: Santalaceae
Chromosome number(s): 2n=30

Description: A shrub *c.* 2 m tall. Leaves sub-sessile, coriaceous, elliptic-oblong, acute, mucronate, glabrous. Male flowers in axillary clusters; female flowers solitary, axillary. Fruit drupe, orange-red, globose.

Distribution: CE; alt. 1100-2600 m.
Himalayas (Himanchal Pradesh to Bhutan), Myanmar, India, Sri Lanka, W. China, Indo-China.

Part(s) used: Heartwood and leaves.

Uses: Heartwood is faintly fragrant and used for adulterating sandalwood. An infusion of leaves has powerful emetic properties. The first infusion in hot water is reddish and nauseating and is discarded; the second, which is used, is yellowish green.

Otochillus porrectus Lindl.

Synonym(s): *Tetrapeltis fragrans* Wall. ex Lindl. **Common name(s):** Jiwanti (Nep.)
Family: Orchidaceae
Chromosome number(s): 2n=4 2n=40

Description: An epiphytic orchid with subcylindric pseudobulbs to 10 cm long. Leaves elliptic, acute OT acuminate. Flowers light yellow in long, flexuous, lax-flowered raceme.

Distribution: WCE; alt. 1900-2300 m.
Himalayas (Uttar Pradesh to Bhutan), NE India, Myanmar, China (Xizang).

Conservation status: Threat unknown. Suspected to be large quantity export.
Part(s) used: Whole plant.

Uses: Used in the treatment of sinusitis and rheumatism. It is also used as tonic.

Oxalis corniculata L.

Synonym(s): *Oxalis pusilla* Salisb.; *O. pusilla* Roxb.; *O. repens* Thunb. **Common name(s):** Chariamilo (Nep.); Paunghayen (Npbh.); Indian sorrel (Eng.)
Family: Oxalidaceae
Chromosome number(s): 2n=24, 28, 44, 48

Description: A small acidic herb, creeping, appressed-pubescent. Leaves trifoliate on long and erect petiole; leaflets obcordate. Inflorescence subumbellate. Flowers yellow with obcordate petals. Fruit a capsule, subcylindric, tomentose.

Distribution: WCE; alt. 300-2900 m. Almost cosmopolitan.

Part(s) used: Roots, leaves and top of the shoots.

Important biochemical constituent(s): Oxalate (Evans 1989). Malic acid in stem and tartaric and citric acids in leaves. Leaves also contain carotene and calcium.

Uses: The plant is used against scurvy. It is a good appetizer. Roots and leaves are used to treat dysentery and diarrhea. Leaf juice is used locally to remove warts and also in the treatment of cataract. Paste of top of the shoots mixed with a few fruits of black pepper is applied to boils, abscesses, wound and weeping eczema.

Paederia foetida L.

Common name(s): Padbiri (Nep.); Prasarini (Sans.)
Family: Rubiaceae
Chromosome number(s): 2n=38, 40-48

Description: A glabrous or puberulous climbing shrub. Leaves opposite, ovate or lanceolate, acute or cuspidate, nerves 4-5 pairs, to 7.5 cm long and 4 cm broad. Flowers violet, shortly pedicelled in slender, trichotomous, often scorpioid cymes. Fruit orbicular, wings pale, 1 cm across.

Distribution: CE; alt. 300-1800 m.
Himalayas C. & E. India, Indo-China, Malaysia.

Part(s) used: Whole plant.

Uses: Roots are emetic. Decoction of the leaves is given in weakness and convalescence due to sickness. Leaf juice is astringent and prescribed for children to control diarrhea. Fruit is used to blacken the teeth by Lepchas believing that it cures toothache. Edible variety of padbiri is considered tonic, diuretic, and aphrodisiac. It is used affliction of stomach and liver. Juice of the roots is given in piles, pains in the chest and the liver and inflammations of spleen. Leaves are used in earache.

Panax pseudo-ginseng Wall.

Synonym(s): *Aralia pseudo-ginseng* Benth. **Common name(s):**
Family: Araliaceae
Chromosome number(s): 2n=24, 48

Description: An erect perennial herb with thick rhizome, to 38 cm tall. Leaves in whorl of 4-6 digitate leaves each with usually 4-6 lanceolate, long-pointed, toothed leaflets. Flowers greenish yellow in umbel borne usually singly on glabrous long-stalk. Fruits globular, berry, c. 8 mm long, shining, red, or half red half black.

Distribution: CE; alt. 2100-2500 m.
Apparently confined to Nepal. But Polunin and Stainton (1985) recorded this plant to occur from Nepal to S.W. China and Burma.

Part(s) used: Root and rhizome

Uses: Considered potent as Korean ginseng. A general tonic, traditionally regarded as a panacea (remedy for troubles, diseases etc.), a magic "cure all" for a whole gallery of complaints ranging from insomnia and toothache to malaria and even epilepsy. In modern Korea ginseng is taken as a fatigue cure.

Pandanus nepalensis Kurz

Synonym(s): *Pandanus diodon* Martelli; *P. furcatus* auct.; *P. furcatus* var. *indica* Kurz
Common name(s): Tarika (Nep.); Screw pine (Eng.)

Family: Pandanaceae

Description: A small tree to 15 m tall with branched aerial roots grown into convoluted masses from the lower part of the stem. Leaves dark green, *c.* 4 m long and 10 cm wide, margin and midrib armed with stout curved spines, arranged spirally like screw around the stem. Flowers unisexual, borne in spikes on several golden-yellow spathes. Fruit resembles pineapple in appearance, orange red when ripe.

Distribution: CE; alt. 700-1000 m. Himalayas.

Part(s) used: Leaves and flowers.

Uses: Leaf extract is bitter and aromatic. Oil extracted from the flowers is said to be antiseptic and also cures headache.

Papaver sominiferum L.

Synonym(s): *Papaver amoenum* Lindl.
Common name(s): Afim (Nep.); Afim swan (Npbh.); Ahifen (Sans.); Opium poppy (Eng.)
Family: Papaveraceae
Chromosome number(s): 2n=20, 22, 44

Description: An erect, rarely branched, usually glaucous annual herb to 120 cm tall. Leaves ovate-oblong or linear-oblong, amplexicaule, lobed, dentate or serrate. Flowers large, with varying colors, usually bluish white with a purple base or white, purple or variegated. Capsule *c.* 2.5 cm diameter, globose, stalked; seeds white or black, reniform.

Distribution: W; alt. 1800 m. S. Europe, C. Asia; cultivated.

Part(s) used: Fruits

Important biochemical constituent(s): Latex of immature fruits contains several alkaloids viz., morphine, codeine, thebaine, noscapine, narceine and papavarine. Other minor alkaloids present in it are aporeine, codamine, cryptopine, gnoscopine, hydrocotarnine, lanthopine, laudanine, laudanosine, meconidine, narcotoline, neopine, papaveramine, prophyroxine, protopine, reticuline, rhoeadine and xanthline. Presence of meconic acid and methyltransferase enzyme has also been reported from this plant (Bhattacharjee 1998).

Uses: Opium alone or in combination with quinine salts, is used in cases of delirium, spasmodic coughing or spasms. It reduces all types of body secretion except for perspiration. It is prescribed to treat diarrhea and diabetes. Of more than two dozen alkaloids contained in this plant, the mostly used ones are morphine, codeine, papaverine, naseopine and thebaine as pain killer, cough depressant, sedative, antitussive and smooth muscle relaxant. Except papaverine, which can also be synthesized, all other alkaloids are obtained either from the latex opium or from direct extraction of capsule.

Paris polyphylla subsp. marmorata (Stearn) H. Hara

Synonym(s): *Paris marmorata* Stearn; *P. violacea* Lév. **Common name(s):** Satuwa (Nep.)
Family: Liliaceae
Chromosome number(s): 2n=10+0-1B

Description: A herb with creeping rhizome with simple, erect stem. Leaves 4-9 in a whorl, petioled, oblong-lanceolate, acuminate. Flowers yellow; sepals 4-6, green, somewhat like the leaf; petals filiform.

Distribution: CE; alt. 2900-3100 m.
Himalayas (Nepal to Bhutan), W. China. *P. polyphylla* Sm.subsp. *polyphylla:* CE; alt. 1800-3300 m. Himalayas Outtar Pradesh to Bhutan), NE India, China. *P. polyphylla* var. *appendiculata* H. Hara: E; alt. 3000-3500 m. *P. polyphylla* var. *walliichii* H. Hara. WCE; alt. 1800-3000 m. Himalayas (Kashmir to Bhutan), NE India, Myanmar, Thailand, W. China.

Conservation status: Vulnerable (IUCN Category). **Part(s) used:** Rhizome.

Important biochemical constituent(s): α-paridine and α-paristyphnin (glucosides).

Uses: It acts as depressant on carotid pressure, myocardium and respiratory movements. It produces vasoconstriction in kidney, but vasodilation in the spleen and limbs and stimulates isolated intestines. The rhizome is anthelmintic and its powder with hot water is used as a tonic.

Parmelia cirrhata Fr.

Synonym(s): *Everniastrum cirrahatum (Fr.)* Hale; *P. cirrhata* var. *divaricata* (Tayl.) Zahlbr.
Common name(s): Jhyau, Charila, Budhna (Nep.); Lichen (Eng.)
Family: Parmeliaceae

Description: A lichen with foliose thallus with involute canaliculate lobes. Lacks soredia and isida. Lower surface lacks rhizinae, if present, in sparse. Spores to 30 µm long and12 µm wide.

Distribution: WCE; alt. 1000-3000 m Himalayas Nepal, India.

Conservation status: Under the Forest Act 1993, His Majesty's Government of Nepal banned on export of Lichen in crude form without processing. However, processed Lichen resinoid is allowed to export.

Part(s) used: Whole plant.

Uses: Used in mental ailments including epilepsy. It is also used in incense sticks, spices and veterinary drugs. The paste is used as ointment and antibiotic in cuts and wound. Lichen resinoid used as a fixative in high-grade perfume.

Parmelia nepalensis (Tayl.) Hook.

Synonym(s): *Everniastrum vexans* (Zahllar.) Hale
Common name(s): Jhyau, Charila, Budhna (Nep.); Lichen (Eng.)
Family: Parmeliaceae

Description: A lichen with foliose thallus with involute canaliculate lobes. Lacks soredia and isidia. Lower surface uniformly rhizinate; rhizinae simple, laciniae, 2-4 mm wide. Spores to 15 µm long and 9 µm wide.

Distribution: WCE; alt. 1000-3000 m Himalayas Nepal, India.
Conservation status: Under the Forest Act 1993, His Majesty's Government of Nepal banned on export of Lichen in crude form without processing. However, processed Lichen resinoid is allowed to export.
Part(s) used: Whole plant.

Uses: Used in mental ailments including epilepsy. It is also used in incense sticks, spices and veterinary drugs. The paste is used as ointment and antibiotic in cuts and wound. Lichen resinoid used as a fixative in high-grade perfume.

Parmelia nilgherrensis Nyl.

Synonym(s): *Parmotrema nilgherrense* (Nyl.) Hale
Common name(s): Jhyau, Charila, Budhna (Nep.); Lichen (Eng.) **Family:** Parmeliaceae

Description: A lichen with foliose thallus, light green above, blakish-brown below, irregularly branched. Lobes large, plane to convoluted, loosely attached to the substratum by rhizinea apothecia. Spores to 20 μm long and 11.5 μm wide.

Distribution: WCE; alt. 1000-3000 m
Himalayas Nepal, India.

Conservation status: Under the Forest Act 1993, His Majesty's Government of Nepal banned on export of Lichen in crude form without processing. However, processed Lichen resinoid is allowed to export.

Part(s) used: Whole plant.

Uses: Used in mental ailments including epilepsy. It is also used in incense sticks, spices and veterinary drugs. The paste is used as ointment and antibiotic in cuts and wound. Lichen resinoid used as a fixative in high-grade perfume.

Parmelia tinctorium Nyl.

Synonym(s): *Parmotrema tinctorum* (Nyl.) Hale
Common name(s): Jhyau, Charila, Budhna (Nep.); Lichen (Eng.)
Family: Parmeliaceae

Description: A lichen with foliose, isidiate, thallus. Thallus lacks marginal cilia, rarely branched; medulla C^+ red cleca noric acid present, apothecia rare, spores 15-18 μm long and 6-9 μm broad.

Distribution: WCE; alt. 1000-3000 m Himalayas Nepal, India.

Conservation status: Under the Forest Act 1993, His Majesty's Government of Nepal banned on export of Lichen in crude form without processing. However, processed Lichen resinoid is allowed to export.

Part(s) used: Whole plant.

Uses: Used in mental ailments including epilepsy. It is also used in incense sticks, spices and veterinary drugs. The paste is used as ointment and antibiotic in cuts and wound. Lichen resinoid used as a fixative in high-grade perfume.

Parnassia nubicola Wall.

Common name(s): Mamira, Nirbisi (Nep.)
Family: Parnassiaceae
Chromosome number(s): 2n=16, 18
Description: A herb with its solitary white flower borne on a slender, 11-30 cm long stem, bearing a single stalkless clasping ovate leaf arising from below the middle of the stem, and with many long-stalked basal leaves having broadly elliptic-heart-shaped, acute blade measuring 2.5-5 cm. Flower 1.5-2.5 cm across, with 5 spreading, often finely toothed, oblong to obovate petals; calyx lobes blunt, hairless; stamens 5, alternating with the 5 fleshy 3-lobed nectaries.
Distribution: WCE; alt. 2900-4200 m.
Afghanistan, Himalayas (Kashmir to Bhutan), China (Xizang).

Part(s) used: Root and rhizome.

Uses: Used in inflammation.

Perilla frutescens (L.) Britton.

Synonym(s): *Perilla ocimoides* Linn.
Common name(s): Ban tulasi (Nep.); Acute common perilla, purple common perilla (Eng.)

Dr. Kamal K. Joshi and Prof. Sanu Devi Joshi

Family: Labiatae
Chromosome number(s): 2n=28, 28+0-2B, 38, 40

Description: An annual herb to 1.5 m tall. Leaves acuminate, pubescent, dentate, crenulate; limb 14 cm long and 6 cm broad; thin and soft leaves with characteristic odor, and the stem and leaves purple in color. Inflorescence an axillary and terminal raceme, to 20 cm long. Flowers small, 3-8, violet. Fruit a collection of globular nutlets, 1.5 mm in diameter, light brown, with pungent taste and aromatic odor.

Distribution: WCE; alt. 600-2400 m.
Himalayas (Kashmir to Bhutan), India, China, Myanmar, Malaysia, Japan (introduced); frequently cultivated.

Part(s) used: Stems, leaves, fruits.

Uses: Stems are useful in feeling of oppression in chest, abdominal distention, morning sickness during pregnancy, fetal distress. Leaves are prescribed in cold, headache, cough, feeling of oppression in chest, nausea, vomiting, food poisoning from fish and crab. Fruits are given in productive cough, and wheezing.

Pholidota articulata Lindl.

Family: Orchidaceae **Chromosome number(s):** 2n=38

Description: An epiphytic 2-leaved herb to 10 cm high with jointed pseudobulbs. Leaves thickly membranous, many-nerved, elliptic, acute, short-stalked, to 10 cm long and 4 cm broad, Inflorescence to 15 cm long, drooping, many-flowered. Flowers to 1.25 cm across, musk-scented and yellowish white (Bhattacharjee 1998).

Distribution: CE; alt. 600-2300 m.
Himalayas (Uttar Pradesh to Sikkim), NE India. Suspected to be exported in large quantity

Part(s) used: Roots and fruits.

Uses: A poultice of root powder is used to treat cancer. Juice of berries is used to treat skin eruptions and cancerous skin ulcers (Bhattacharjee 1998).

Phyllanthus amarus Schumacher & Thonn.

Synonym(s): *Phyllanthus niruri* auct.
Common name(s): Bhuin amala, Bhumyalaki (Nep.)
Family: Euphorbiaceae
Chromosome number(s): 2n=14, 26, 28

Description: A small glabrous pale green herb. Leaves variable, sub-sessile, stipule subulate, small, membranous, elliptic - obovate, oblong or linear, tip rounded, obtuse or acute, nerves obscure. Flower numerous, shortly pedicelled. Male flower solitary, 2-nate, almost sessile. Sepals 5-6, oblong, enlarged in fruit. Style 3, minute. Fruit an ovoid or globose capsule.

Distribution: CE; 470-900 m. Pantropical.

Part(s) used: Whole plant.

Important biochemical constituent(s): Three crystalline lignans including phyllanthine and hypophyllanthine have been isolated from the aerial parts (Anonymous 1994).

Uses: Acrid, cooling, and useful in thirst, bronchitis, leprosy, anemia, urinary discharges, biliousness, asthama, and hicough. The plant is also used as a diuretic and in menorrhogia.

Phyllanthus emblica L.

Synonym(s): *Phyllanthus taxifolius* D. Don; *Emblica officinalis* Gaertn.
Common name(s): Amala (Nep.); Ambah (Npbh.); Amalaki (Sans.); Gooseberry (Eng.)
Family: Euphorbiaceae
Chromosome number(s): 2n = 28, 98, 104, 196

Description: A deciduous tree up to 10 m high. Branchlets glabrous or finely pubescent. Leaves distichously set giving the false appearance of a pinnate leaf, subsessile stipule scarious, linear, oblong, obtuse, entire, and glabrous, up to 1.5 cm long. Flowers densely fascicled, yellow. Male flower long-pedicelled. Sepals 5-6, obovate-oblong. Stamens 3; filaments united into a short column; anther erect, free, disc absent. Female flower: sepals as in male flowers. Style 3, connate below. Stigmas 2-fid, dilated. Fruits a globose capsule, 2 cm in diameter, six-lobed, pale yellow or greenish (Malla *et al.* 1986).

Distribution: WCE; alt. 150-1400 m.
Himalayas (Uttar Pradesh to Bhutan), NE India, N. Myanmar, S. China, Malaysia.

Part(s) used: Fruits.

Important biochemical constituent(s): Trigalloyl glucose, terchebin, corilogin and ellagic acid, tannin and polyphenolic compounds, phyllantidine, and phyllantine (Rajbhandari *et al.* 1995).

Uses: Fruit - acrid, cooling, refrigerant, diuretic and laxative. Raw fruit - aperient. Dry fruit - useful in haemorrhage, diarrhea and dysentery, in combination with iron used in anaemia, jaundice and dyspepsia (Rajbhandari *et al.* 1995). Fruit poultice is used to stop bleeding from cuts, fruit powder as coolant and laxative, fruit juice in indigestion, jaundice, anaemia and heart complaint. Fruits are good liver tonic. It is a rich source of vitamin C. Seeds are administered against asthma, stomach disorder and bronchitis. Its fruit is one of the three ingredients of the famous ayurvedic preparations "Triphala" and "Chyawanprash".

Phyllanthus urinaria L.

Common name(s): Bhui amala, Bhumyalaki (Nep.); Amalaki (Sans.)
Family: Euphorbiaceae
Chromosome number(s): 2n=14, 52

Description: An annual herb branched diffusely. Leaves distichously imbricate, oblong or linear oblong. Flowers minute, yellowish. Capsule globose, scarcely lobed, echinate. Seeds transversely furrowed.

Distribution: WCE; alt. 760-1700 m. Pantropical.

Part(s) used: Whole plant.

Uses: It is an excellent diuretic. The juice of leaves is an appetizer to children.

Phytolacca acinosa Roxb.

Synonym(s): *Phytolacca latbenia* (Moq.) H. Walter
Common name(s): Jaringo (Nep.); Sweet belladonna (Eng.)
Family: Phytolaccaceae
Chromosome number(s): 2n=18, 36

Description: A tall herb on rocks with succulent stem. Leaves petioled, elliptic-ovate or elliptic, acute or acuminate, membranous. Flowers white, in erect, long, many-flowered, racemes; bracts linear-lanceolate. Fruit ovoid, dark purple, juicy.

Distribution: WC; alt. 2200-3200 m.
Himalayas (Kashmir to Arunachal Pradesh), NE India, Laos, W. China.

Part(s) used: Whole plant.

Important biochemical constituent(s): Phytolacca-toxin in seeds.

Uses: The plant is believed to have narcotic effect. The fruit is occasionally eaten and used as flavoring agent.

Piper longum L.

Common name(s): Pipala, Murjhang, Pipalamul (Nep.); Pipee (Npbh.); Pippali (Sans.); Long pepper (Eng.)
Family: Piperaceae
Chromosome number(s): 2n=24, 26, 48, 52, 96

Description: A creeping aromatic herb. Lower leaves ovate, cordate with big lobes at the base to 10 cm long; upper leaves dark green, oblong-ovate, cordate. Flowers in spikes, bracts of male spikes narrow and those of female spikes circular. Fruits small, ovoid, shining, blackish green.

Distribution: WCE; alt. 200-800 m.
Himalayas (Nepal to Bhutan), India, Sri Lanka, Malaysia.

Part(s) used: Roots and fruits.

Important biochemical constituent(s): Piperine and piplartine. Dried fruits yield 0.7% of essential oil with spicy odor (Bhattacharjee 1998).

Uses: Roots and fruits are used to treat insomnia, epilepsy, obstruction of bile duct and gall bladder, dysentery and leprosy. Dried unripe fruits are given in cold, cough, chronic bronchitis and diarhhoea. It is also used in liniments for rheumatic pains and paralysis. It is anthelmintic and carminative.

Pistacea chinensis Subsp. integerrima J. L. Stewart Rech. F.

Synonym(s): *Pistacia integerrima* J. L. Stewart
Common name(s): Kakad singhi (Nep.) for insect gall on *Pistacea chinensis* Subsp. *integerrima* J. L. Stewart and *P. khinjuk* Stocks.
Family: Anacardiaceae
Chromosome number(s): 2n=30

Description: A deciduous tree to 18 m tall and 2.7 m in girth with a short stout bole. Bark gray or blackish, viscous or aromatic when cut. Leaves pinnate; leaflets lanceolate, 7-12 cm long. Flowers in panicles, small, reddish. Fruit a globose drupe, *c.* 6 mm diameter, rugose, gray when ripe. Insect galls are produced on leaves and petioles. These galls are pale greenish brown in color, horn-shaped, hard, rugose and hollow, varying in length mostly of about 3.8 cm but sometimes to 30 cm.

Distribution: W; alt. 2100 m.
Afghanistan, Pakistan (Chitral), Himalayas (Punjab, Nepal).

Conservation status: Rare (IUCN Category).

Part(s) used: Insect galls produced on the leaves and the petioles.

Important biochemical constituent(s): Plant contains pistacenenoic acid A & B. Gall contains tannins, an essential oil and a resin (Anonymous 1994).

Uses: Tonic, expectorant, used in fever, asthma etc. Powdered galls fried in ghee and given orally in dysentery.

Plectranthus mollis (Aiton) Spreng.

Synonym(s): *Plectranthus cordifolia* D. Don; *Plectranthus incanus* Link.
Family: Labiatae
Chromosome number(s): 2n=28

Description: An erect herb or undershrub to 120 cm tall. Leaves ovate-cordate, acute or acuminate, crenate. Flowers pale lilac or light blue, in lax-flowered cymes in racemes. Nutlets subglobose, smooth, pale brown, dotted with dark spots.

Distribution: CE; alt. 900-1500 m.
Himalayas (Uttar Pradesh to Sikkim), India, Sri Lanka.

Part(s) used: Leaves and flowering tops.

Important biochemical constituent(s): Essential oil, resin and tannin.

Uses: The oil possesses antibacterial activities. It also acts as cardiac depressant, respiratory stimulant, vaso-constrictor and relaxant on smooth and skeletal muscles.

Plumbago zeylanica L.

Synonym(s): *Plumbago rosea* L.
Common name(s): Chitu (Nep.); Chitraka (Sans.); White flower leadwort (Eng.)
Family: Plumbaginaceae
Chromosome number(s): 2n=16, 28

Description: A perennial, sub-scandent shrub. Leaves ovate, glabrous. Flowers white, in elongated spikes. Capsules oblong, pointed, contained in viscid glandular persistent calyx.

Distribution: WCE; alt. 100-1300 m. Tropical Africa, tropical Asia.

Part(s) used: Roots.

Important biochemical constituent(s): Root contains an acrid crystalline principle plumbagin. It also contains chloroplumbagin and biplumbagin (Anonymous 1994).

Uses: Root paste is applied as antibiotic on abscess. Roots are used to promote appetite and improve digestion. It is also used to treat toothache, piles, and diarrhea. The plant is acrid, produces irritation and rubefaction when applied to skin.

Podophyllum hexandrum Royle

Synonym(s): *Podophyllum emodi* Wall. ex Hook. f. and Thoms.; *P. emodi* var. *hexandrum* (Royle) Chatterjee & Mukerjee; *P. hexandrum* var. *bhootanense* (Chatterjee & Mukerjee) Browicz; *P. hexandrum* var. *emodi* (Hook. f. & Thomson) Seliv.-Gor.
Common name(s): Papra (Nep.); Meme gudruk (Dolp.); Laghupatra (Sans.); Indian podophyllum (Eng.)
Family: Berberidaceae
Chromosome number(s): 2n=12, 12+0-2f

Description: An erect, glabrous, succulent herb to 66 cm tall, with creeping, perennial rhizome bearing numerous roots. Leaves 2 or 3, orbicular-reniform, palmate, peltate, with lobed segments. Flowers solitary, white or pink, cup-shaped. Fruit an oblong or elliptic berry to 5 cm diameter, orange red. Seeds many embedded in the pulp.

Distribution: WCE; alt. 3000-4500 m.
Afghanistan, Himalayas (Kashmir to Arunachal Pradesh), W. China.

Conservation status: Vulnerable (IUCN Category).
Part(s) used: Roots and rhizome.

Important biochemical constituent(s): Podophyllin.

Uses: Root paste is applied on ulcer, cuts and wound. It is also used as purgative and in the treatment of skin diseases and the growth of tumor.

Polypodium vulgare L.

Common name(s): Bisfez (Nep.); Wall fern (Eng.)
Family: Polypodiaceae
Chromosome number(s): 2n=74, 148

Description: An epiphytic fern with creeping, cylindric rhizome 4.5-10.5 cm long, 2-6 cm wide, grayish black, green internally. Fronds stipitate, free veins, sori dorsal on the frond.

Distribution: C; alt. 2400-2700 m.
Part(s) used: Rhizome, fronds.

Important biochemical constituent(s): Triterpenes

Uses: Aperient (Evans 1989). Rhizome is used as purgative and in the treatment of cholagogue.

Potentilla josephiana H. Ikeda & H. Ohba

Synonym(s): *Potentilla fulgens* var. *intermedia* Hook. f.; *Potentilla lineata* var. *intermedia* (Hook. f.) S. N. Dixit & Panigrahi
Common name(s): Bajradanti (Nep.)
Family: Rosaceae
Chromosome number(s): 2n = 28, 34, 56.

Description: An erect softly silky herb. Leaves alternate, stipulate, pinnately compound, 9-20 cm long, leaflets numerous pairs alternately large and small diminishing in size from the uppermost pairs downwards, sessile, terminal leaflets 3-4.5 cm long and 1.5-2 cm broad, ovate or oblong-ovate, obtuse, closely and sharply toothed, upper surface green, hairy, lower silvery tomentose. Inflorescence terminal corymbs, 8 cm long and 5 cm broad. Flowers pedicelled, bracteate, yellow. Calyx bowl-shaped, silvery, hairy. Petals 5, free, ovate, obtuse, yellow. Stamens numerous; filaments 0.4 cm long; anther traingular, bilobed, dorsifixed, yellow. Carpels numerous, superior, crowded on a small dry receptacle. Style lateral. Stigma simple. Achenes glabrous (Malla *et al.* 1986).

Distribution: WCE; alt. 2400-4150 m.
Himalayas (Uttar Pradesh to Sikkim), China (Xizang, Yunnan).

Part(s) used: Root.

Important biochemical constituent(s): Flavonoids, polyphenol, sterols, carotenes, coumarin and 2.59% tannin (Rajbhandari *et al.* 1995).

Uses: The rootstocks are reported to be used in diarrhea. Root powder is used for toothache and as tooth powder for cleaning teeth.

Prunella vulgaris L.

Common name(s): Selfheal, Heal-all, Sicklewort (Eng.) **Family:** Labiatae
Chromosome number(s): 2n=28, 32

Description: A small low sprawling, faintly pubescent perennial herb to 45 cm tall, creeping at the base. Leaves stalked, upper ones sessile, ovate or oblong, entire, toothed or pinnatifid. Flowers dimorphic, larger bisexual and smaller female with purplish corolla. Fruit a collection of ovoid nutlets, smooth, 2-celled when young, splitting into 4 parts when mature, each part containing 1 seed

Distribution: WCE; alt. 1200-3800 m. Throughout Europe and temperate Asia.

Part(s) used: Inflorescence or whole plant.

Uses: It is used in dizziness due to hypertension, headache, tinnitusand conjuctivitis. It is also given in cough, skin inflammation and boils.

Prunus armeniaca L.

Common name(s): Khurpani (Nep.); Khurpani (Npbh.); Common apricot (Eng.)
Family: Rosaceae
Chromosome number(s): 2n=16

Description: A tree *c.* 10 m tall with reddish bark. Leaves ovate to round-ovate or some times subcordate, to 9 cm long. Flowers pinkish white, borne singly and appearing in advance of the foliage. Fruits round, c. 5 cm across, pubescent when young, but nearly glabrous at maturity, with a yellow skin overlaid with red; flesh yellow to yellowish orange, firm and sweet, mostly free from the flat, ridged stone; kernels sweet or bitter types available.

Distribution: C; alt. 2900-3500 m.
China; widely cultivated, sometimes escaped cultivation.

Part(s) used: Kernel oil.

Uses: The oil is used in earache and other ailments.

Prunus creasoides D. Don

Synonym(s): *Cerassus puddum* Seringe; *Prunus puddum* Roxb. ex Brandis
Common name(s): Paiyun (Nep.); Padmaka (Sans.); Himalayan wild cherry (Eng.)
Family: Rosaceae
Chromosome number(s): 2n=16

Description: A medium-sized more or less glabrous tree. Leaves simple, ovate or oblong-lanceolate, caudate, acuminate, sharply serrate; petiole with 2-4 glands. Flowers rose-red or whitish, solitary, fascicled or umbelled. Fruit a drupe, oblong or elliptical, stone rugose and furrowed.

Distribution: WCE; alt. 1300-2400 m.
Himalayas (Punjab to Bhutan), NE India, Myanmar, W. China.

Part(s) used: Smaller branches.

Uses: Smaller branches are crushed and soaked in water and taken internally to stop abortion.

Prunus persica (L.) Batsch.

Common name(s): Aru (Nep.); Bäsi (Npbh.); Peach (Eng.) **Family:** Rosaceae
Chromosome number(s): 2n=16

Description: A small tree to 8 m tall, with glabrous twigs. Leaves oblong to broad lanceolate, serrate, glabrous. Flowers solitary, pink. Fruits subglobose, 5-7 cm across, fleshy with a hard and deeply pitted stone.

Distribution: W (Juphal, Dolpa); alt. 2000 (Ghimire *et al.* (2000) China; widely cultivated in Asia and Europe.

Part(s) used: Leaves

Uses: The leaves are laxative and were formerly used as anthelmintic. An infusion of leaves and bark is given in coughs, specially whooping cough. The leaves, blossoms and the kernels are poisonous. The flowers are also used as purgative and anthelmintic.

Pueraria tuberosa (Roxb. ex Willd.) DC.

Common name(s): Baralikand (Nep.); Vidari (Sans.); Indian kudzu (Eng.)
Family: Leguminosae
Chromosome number(s): $2n = 22$

Description: A wisteria-like deciduous woody climber, with trifoliate leaves having large leaflets. Leaves long-stalked; leaflets 10-20 cm, silky-haired when young and remaining thinly hairy beneath, the terminal broadly ovate long pointed, the lateral leaflets unequal-sided. Spikes dense with numerous bright mauvish-blue flowers appearing when the plant is leafless. Spikes 15-40 cm long, on axillary branches; petals 1-1.8 cm; calyx densely silky. Pod flat, constricted between seeds, covered with bristly brown hairs, 5-8 cm (Polunin and Stainton 1986).

Distribution: WC; alt. 300-1500 m.
Tropical Himalayas (Kashmir to Nepal), India.

Part(s) used: Tuberous root.

Important biochemical constituent(s): ß-sitosterol and pterocarpan tuberosia.

Uses: Cooling, aphrodisiac, emetic, tonic, lactogogue, diuretic, alterative, clears the voice, cures biliousness, burning sensation, urinary discharges.

Punica granetum L.

Common name(s): Darim (Nep.); Dhale (Npbh.); Pomgranate (Eng.)
Family: Punicaceae
Chromosome number(s): 2n=14-16, 18

Description: A deciduous shrub or sometimes a small tree with thin smooth gray bark. Leaves to 8 cm long, entire, lanceolate to broadly oblanceolate, opposite, shining. Flowers scarlet with crinkle petals, 4-5 cm across and with large scarlet or brownish orange-like fruits; calyx scarlet, 2-3 cm long, tubular, with 5-7 triangular fleshy lobes; petals 5-7, scarlet, broadly obovate, 1.5-3 cm; stamens numerous. Fruits globular, 4-8 cm, crowned with calyx and with a thick leathery rind, flesh pink juicy with white, pink or red seeds.

Distribution: WC; alt. 700-2700 m.
C. & W. Asia, Himalayas; widely cultivated in E. Himalayas, S. Europe, Asia.

Part(s) used: Roots, bark, fruit pulp and seeds.

Important biochemical constituent(s): Roots and bark contain tannin (20-22%), and alkaloids (0.5-1%). Seeds contain steroidal oestrogen. The fruit pulp contains protein, carbohydrate, fat, fibre, minerals, oxalic acid and vitamins A, B and C.

Uses: Tannin is obtained from root, bark, stem, leaves and fruit rind. Bark is reported to be very effective anthelmintic to expel tapeworms. Bark and the rind of fruits are used to treat dysentery and diarrhea. Fruit pulp is beneficial in cardiac disorders and the fruit juice is prescribed for leprosy patients. Fruit pulp and seeds are stomachic.

Quisqualis indica L.

Common name(s): Rangoon creeper (Eng.) **Family:** Combretaceae
Chromosome number(s): 2n=22, 24, 26

Description: A large woody scandant shrub. Leaves opposite or sub-opposite, papyraceous, elliptic, or elliptic oblong, acuminate, entire. Flowers numerous, white or red, fragrant, in axillary or terminal pendulous racemes. Fruit dry, coriaceous, ovate-elliptic, 2.5-4.0 cm long and 0.75-1.25 cm broad, 5-angled or 5-winged and 1-seeded.

Dr. Kamal K. Joshi and Prof. Sanu Devi Joshi

Distribution: WE; alt. 200-600 m. Old World tropics.

Part(s) used: Whole plant.

Important biochemical constituent(s): An active principle resembling santonin.

Uses: Extracts of roots and leaves are used as anthelmintic. Leaves are also used in a compound decoction to relieve flatulence. Fruits and seeds possess better anthelmintic properties. Fruits are picked half-ripe when they are bitter, pulped in water and the liquid taken internally. Seeds from ripe fruits may also be used only after strict moderation. More than 4 or 5 seeds are reported to cause colic. An Overdose of seeds causes unconsciousness.

Ramalina Ach. sps.

Common name(s): Jhyau, Charila, Budhna (Nep.); Lichen (Eng.)
Family: Usneaceae

Description: A lichen with fruticose, erect to procumbent, branched thallus. Branches rounded to strap-shaped or flattened lobes, corticated both sides; green alga as photobiont. Apothecia lecanorine; asci 8-spored, hyaline, 2-celled, straight or slightly curved.

Distribution: WCE; alt. *c.* 3000 m Himalayas Nepal, India.

Conservation status: Under the Forest Act 1993, His Majesty's Government of Nepal banned on export of Lichen in crude form without processing. However, processed Lichen resinoid is allowed to export.

Part(s) used: Whole plant.

Uses: Used in mental ailments including epilepsy. It is also used in incense sticks, spices and veterinary drugs. The paste is used as ointment and antibiotic in cuts and wound. Lichen resinoid used as a fixative in high-grade perfume.

Rauvolfia serpentina (L.) Benth. ex Kurz.

Common name(s): Sarpagandha (Nep.); Sarpagandha (Sans.); Rauvolfia, Serpent wood (Eng.)
Family: Apocynaceae
Chromosome number(s): 2n=20, 22, 24, 44

Description: An erect, glabrous, perennial shrub attaining to 75 cm in height. Leaves arranged in whorls of 3 to 4, to 10 cm long and 5 cm broad, gradually tapering into a short petiole, bright green above and pale beneath. Flowers white or pinkish, with deep red peduncle, around 1.5 cm long, arranged in small clusters. Fruits round, to 5 cm in diameter, dark, purple or blackish when ripe.

Distribution: CE; alt. 100-900 m.
Tropical Himalayas, India, Sri Lanka, Malaya.

Conservation status: Endangered (IUCN Category). **Part(s) used:** Root bark.

Important biochemical constituent(s): Of more than three dozen alkaloids contained in the root bark, the most common ones are reserpine, rescinnamine and deserpidine.
Uses: It has anti-hypertension and sedative properties. It is, therefore, used to treat high blood pressure and as tranquilizer. Roots are also useful in bowel disorder and during fever.

Rhamnus napalensis (Wall.) M.A. Lawson

Synonym(s): *Ceanothus nepalensis* Wall. **Common name(s):** Chile kath (Nep.)
Family: Rhamnaceae
Chromosome number(s): 2n=24, 34

Description: An unarmed shrub. Leaves petiolate, oblong or elliptic-oblong, shortly acuminate. Flowers green, in racemes. Fruit obovate, red.

Distribution: CE; alt. 600-1700 m
Himalayas (Nepal to Bhutan), NE India, N. Myanmar, Indo-China, China, Malaysia.

Part(s) used: Fruits.
Uses: Fruit pounded and macerated in vinegar is prescribed in herpes.

Dr. Kamal K. Joshi and Prof. Sanu Devi Joshi

Rheum australe D. Don

Synonym(s): *Rheum emodi* Wall. ex Meissner
Common name(s): Padamchal (Nep.)
Family: Polygonaceae
Chromosome number(s): 2n=22, 44

Description: A herb to 2 m tall with very stout rootstock and green and brown streaked stem. Leaves with a very stout leaf stalk, 30-45 cm long and with rounded to broadly ovate blade with a heart-shaped base, hairy beneath, the basal leaves to 60 cm across. Flowers small, 3 mm across, dark reddish purple borne in dense branched clusters, to 30 cm long. Nutlets purple, *c.* 12 mm long, with narrow wings, and with a rounded heart-shaped base and notched apex.

Distribution: CE; alt. 3200-4200 m.
Himalayas (Himachal Pradesh to Nepal, ?Bhutan), China (Xizang).

Part(s) used: Root

Uses: Paste of the root is taken orally with **Curcuma** for curing internal injury and applied on forehead for the relief from headache.

Rheum nobile Hook. f. & Thomson

Common name(s): Padamchal (Nep.)
Family: Polygonaceae

Description: A very striking and distinctive-looking herb with a stout erect stem bearing a slender conical spike of large pale cream-colored, rounded and bladder-like drooping and overlapping bracts which conceal the short flower clusters. Stem very stout to 1.5m. Leaves leathery, to 30 cm across, rounded with a wedge-shaped or rounded base, margin usually edged with red, and with a stout stalk; leaves gradually changing upwards into the bracts. Bracts progressively smaller up stem, the lowest to 15 cm and the uppermost 1-2 cm. Flowers-clusters branched, to 6 cm; flowers green, very numerous, *c.* 2 mm across. Nutlets broadly 2-4-winged.

Distribution: E; alt. 3900-4300 m. Himalayas (Nepal to Bhutan).

Conservation status: Rare (IUCN Category).

Part(s) used: Roots.

Uses: Watery extract of the root is taken orally in stomach pain, constipation, dysentery, swelling of throat and tonsilitis. Lotion drops are used in ears to relieve from earache (Bhattacharjee 1998).

Rhododendron anthopogon D. Don var. anthopogan

Common name(s): Sunpati (Nep.) **Family:** Ericaceae
Chromosome number(s): 2n=26

Description: A small strongly aromatic shrub to 60 cm tall with brown scaly young shoots. Leaves oval to obovate, 2.5-4.0 cm long, densely scaly beneath. Flowers *c.* 2 cm across, white or yellow tinged with pink, in compact clusters of 4-6; corolla with a narrow tube and 5 rounded spreading lobes. Capsule c. 3 mm encircled by the persistent calyx. (var. *hypenanthum* Balf. f.) Hara has yellow flowers and winter bud-scales persisting for several years.

Distribution: WCE; alt. 3300-5100 m.
Himalayas (?Punjab to Bhutan), China (Xizang). *R. anthopogan* var. *hypenanthum* (Balf. f.) H. Hara: WCE; alt. 3300-5000 m. Himalayas (Kashmir to Bhutan).

Part(s) used: Leaves and flowers.

Uses: Decoction of leaves is used in cold, cough and chronic bronchitis. Fragrant dried and powdered flowers mixed with bland oil are used for massage over the entire body in post-delivery complications like fevers, cough and cold (Bhattacharjee 1998). Fragrant dried flowers are used to drink as tea by local people.

Rhododendron arboreum Sm.

Synonym(s): *Rhododendron puniceum* Roxb.
Common name(s): Lali gurans (Nep.); Rhododendron (Eng.)
Family: Ericaceae

Dr. Kamal K. Joshi and Prof. Sanu Devi Joshi

Chromosome number(s): 2n=26

Description: A small to medium-sized evergreen tree with soft, red-brown to gray bark. Leaves crowded at the end of the branches, *c.* 12 cm long, narrow, pointed at both ends, leathery, shiny, silvery beneath. Flower large, attractive, bell-shaped, in bunches at the ends of the branchlets, generally a deep crimson. Fruit a capsule, 2-3 cm long, and *c.* 1 cm diam., oblong, slightly curved, and ridged lengthwise; seeds numerous, dark brown, compressed, oblong with tuft of hairs at both ends.

Distribution: Rhododendron arboreum Sm. var. **arboreum.** WCE; alt. 1500-3300 m.
Himalayas (Kashmir to Arunachal Pradesh), China (Xizang), NE India, Myanmar.

Part(s) used: Flowers.

Uses: Used for curing dysentery and a paste made of flower is said to bring relief from headache.

Rhododendron lepidotum Wall. ex G.Don

Synonym(s): *Rhododendron salignum* Hook. f.
Common name(s): Bhale-sunpati (Nep.); Belu (Dolp.)
Family: Ericaceae
Chromosome number(s): 2n=26

Description: A resinous shrub to 1 m, very variable in size, habit and flower color. Flowers in clusters of usually 2-4, slender-stalked pink, dull purple, or pale yellow in color, each 2-2.5 cm across. Corolla shortly and broadly tubular, with 5 spreading rounded lobes, scaly and glandular outside; stamens exserted, 8, filaments hairy below. Leaves small, narrow-oblanceolate 1.2-4 cm, scaly above and beneath. Capsule *c.* 2 cm, densely scaly (Polunin and Stainton 1986).

Distribution: WCE; alt. 2100-4700 m.
(Kashmir to Arunachal Pradesh), N. Myanmar, W. & S. China. **Part(s) used:** Bark and leaves.

Uses: Bark taken as tea is supposed to be purgative (Bhattacharjee 1998). Leaves are aromatic, used in incense, stimulant. Essential oil obtained from leaves are used in high grade perfume.

Rhynchostylis retusa (L.) Blume

Family: Orchidaceae **Chromosome number(s):** 2n=38

Description: An epiphytic herb attaining 60 cm high with stout and woody stems clothed with sheaths of fallen leaves. Leaves leathery, strap-shaped, linear, to 30 cm long and 2.5 cm broad. Inflorescence pendulous, compact, many-flowered, to 60 cm long. Flowers 1 cm across, fragrant, white, spotted with bluish purple.

Distribution: CE; alt. 1200-1500 m.
Himalayas (Nepal, Sikkim), India, Sri Lanka, Myanmar.

Conservation status: Commercially threatened (IUCN Category). **Part(s) used:** Leaves.

Uses: Leaves are used to treat rheumatism. Fresh plant is used as an emollient.

Ricinus communis L.

Common name(s): Ander, Arandi, Eranda (Nep.); Aleha (Npbh.); Eranda (Sans.); Castor bean (Eng.)
Family: Euphobiaceae
Chromosome number(s): 2n = 20

Description: A herbaceous plant or a soft-wooded shrub readily recognized by its large green or reddish, long-stalked, palmately 5-9-lobed leaves 20-50 cm across. Flowers reddish or yellowish, in terminal spikes to 15 cm; the upper flowers female, the lower male. Leaf-blade with triangular-lanceolate toothed lobes cut to two-thirds the width of the blade; leaf stalk as long as the blade. Fruit distinctive, in a cylindric cluster of many capsules, each 2.5 cm, usually covered with long stiff prickles, splitting into 3 valves, and with large smooth shiny mottled gray, white or brown seeds, each with a large swelling (aril) at its base (Polunin and Stainton 1986).

Distribution: WCE; alt. 150-2400 m.
Believed to be native of NE tropical Africa; widely cultivated and occasionally naturalized throughout the tropics (Press *et al.* 2000).

Part(s) used: Root bark, leaves, flowers, seeds and oil.

Important biochemical constituent(s): Castor seeds constitute glycoprotein, ricin, ricinnie and lipase. Small seeds are rich in oil than the larger ones (Bhattacharjee 1998).

Uses: Used in, pains, ascites, fever, asthma, bronchitis, and externally in skin inflammation and leprosy. Leaves are used in intestinal worms, night-blindness etc. Flowers are used in glandular tumors. Seeds and oil are cathartic and aphrodisiac and used in lumbago, leprosy, constipation, etc. Oil is also used as tonic, emollient, and laxative. It is prescribed in cases of inflammation of the intestine, dysentery and skin diseases. Castor oil is also used in contraceptive jelly, foam and cream. Oil, root bark and pounded leaves are used as purgative (Bhattacharjee 1998).

Rosa laevigata Michx.

Synonym(s): *Rosa sinica* auct.
Common name(s): Cherokee rose (Eng.)
Family: Rosaceae
Chromosome number(s): 2n=14

Description: An evergreen climbing shrub, c. 5 m tall. Stem with a few compressed somewhat hooked prickles. Leaves pinnate; leaflets 3-5, oval-elliptical or oval-lanceolate, 2-7 cm long and 1.5-4-5 cm broad, acute or acuminate, finely dentate, the terminal leaflet larger and long-petiolate. Flowers solitary, fragrant, white, 6-8 cm in diameter; calyx a cupuliform tube covered with hairs; petals cordate. Receptacle oblong, ovoid, or nearly globular, to 3 cm long and 1.5 cm broad, reddish, covered with stiff hairs and crowned with persistent calyx.

Distribution: C; alt. 1200 m.
China; often cultivated in N. India and Japan.

Part(s) used: Fruit.

Uses: Useful in malabsorption and diarrhea, chronic cough, perspiration from no apparent cause (night sweats). Also used in enuresis (frequent micturition), leucorrhoea (massive uterine bleeding), and seminal emission.

Rubia manjith Roxb. ex Fleming

Synonym(s): *Rubia cordifolia* auct.; *R. cordifolia* forma *rubra* Kitam.; *R. cordifolia* var. *khasiana* Watt.; *R. cordifolia* var. *mungith* Roxb. ex desv.; *R. munjista* Roxb.; *R. wallichiana* Decne.
Common name(s): Majitho (Nep.); Manjistha (Sans.); Indian madder (Eng.)
Family: Rubiaceae
Chromosome number(s): 2n=22, 44, 66 and 132

Description: A perennial climbing herb, with 4-angled stems and branches having hooked prickles on angles. Leaves in whorls of 4, ovate-heart-shaped long-pointed blade, 3-5 cm with hooked prickles on the veins beneath, leaf-stalk as long as blade with hooked prickles. Flowers 3 mm across, with 5 spreading incurved lobes, reddish brown, in small clusters aggregated together into a large branched cluster with small leafy bracts. Fruits *c.* 5 mm, globular, black, fleshy, with red juice.

Distribution: CE; alt. 1200-3200 m
Himalayas (Himachal Pradesh to Bhutan), NE India (Meghalaya).

Part(s) used: Roots.

Important biochemical constituent(s): Septilin. Also contains sterol, terpene and saponin.

Uses: Roots have astringent, antidysenteric, antiseptic and deobstruent properties. They are used as tonic and also to cure rheumatism. Roots form an ingredient of several Ayurvedic preparations. They are believed to be effective against Staphylococcus aureus and their paste is applied in ulcers, inflammations and skin diseases. A decoction of stems and leaves is used as a vermifuge.

Saccharum spontaneum L.

Synonym(s): *Saccharum canaliculatum* Roxb.; *S. semidecumbens* Roxb.
Common name(s): Kasa (Sans.)
Family: Gramineae
Chromosome number(s): 2n=40-128 (40, 48, 50, 52, 54, 56, 58, 60, 62, 64, 66, 68, 70, 72, 76, 78, 80, 82, 88, 90, 92, 94, 96, 100, 104, 112, 116, 124, 126 and 128.

Description: A perennial grass with slender culms. Culms green, gray, ivory or white, hard, but very pithy, and often hollow in the center, 5-15 mm in diameter, often rooting at the nodes; internodes usually long and nodes always thicker than the internodes. Leaves long, linear, narrow or very narrow and sometimes reduced to the midrib; the ratio of breadth to length ranges from 1 : 24 to 1 : 300 or more in the different forms of the species. Inflorescence a panicle, pale or grayish white to purplish gray. Spikelets in pairs, one pedicelled and the other sessile; the pedicelled spikelet of the pair always blooming first; glumes always four and lodicules ciliate.

Distribution: WCE; alt. 200-1700 m. Warmer regions of Old World.

Part(s) used: Roots.

Uses: Decoction of roots is useful in inflammations, thirst, haemostatics, dysuria and urolithiasis. Treatment in the form of cold infusion can also be given in some these ailments. In inflammation of hands and legs, paste of roots is applied locally.

Sapindus mukorossi Gaertn.

Common name(s): Ritha (Nep.); Hathan (Npbh.); Phenila, Urista (Sans.); Soapnut tree (Eng.)
Family: Sapindaceae
Chromosome number(s): 2n=36

Description: A deciduous tree, 18 m or more in height and c. 1.8 m in girth. Bark dark-greenish gray, pale-gray or brown, fairly smooth with vertical lines of lenticels and fine fissures exfoliating in irregular flakes. Leaves paripinnate, crowded near the ends of the branches; leaflets lanceolate, to 15 cm long and 5-10 pairs. Flowers polygamous, mostly bisexual, small, in terminal compound panicles. Fruit a drupe, globose, fleshy, saponaceous, usually solitary, sometimes two drupels together, wrinkled or smooth, 2.0-2.5 cm in diameter; dried pulp is light brown, somewhat translucent, saponaceous rind with a wrinkled surface. Seed enclosed in black, smooth, hard endocarp.

Distribution: WE; alt. 1000-1200 m.
Himalayas, NE India, Myanmar, Indo-China, China, Taiwan, Korea, Japan.

Part(s) used: Fruits.

Important biochemical constituent(s): Fruit pericarp contains sesquiterpene, oligoglycosides, mukuroziosides and large amount of saponins (Anonymous 1994).

Uses: Fruits are believed to have expectorant and emetic properties and are used in excessive salvation, epilepsy and chlorosis. Powdered seeds possess insecticidal properties. They are given in dental caries.

Saraca asoca (Roxb.) W.J. de Wilde

Synonym(s): *Saraca indica auct.*
Common name(s): Ashoka (Nep.); Ashoka (Npbh.); Ashoka (Sans.); Ashok tree (Eng.)
Family: Leguminosae
Chromosome number(s): 2n = 24

Description: An evergreen tree of 6-10 m high with compound leaves; leaflets linear-lanceolate, glabrous having tough texture and upper surface shining. Flowers in cluster, petals golden yellow and anthers red. Fruit a pod, compressed, 9-23 cm long with 4-8 compressed seeds (Adhikari 1998).

Distribution: CE; alt. 150-1400 m.
Himalayas (Uttar Pradesh to Nepal), India, Sri Lanka, Myanmar.

Part(s) used: Bark and flowers.

Important biochemical constituent(s): Tannins and catechin, an organic compound of calcium, haemotoxyline and ketosterol.

Uses: The bark is bitter and acrid; used as refrigerant, astringent, anthemintic, demulcent, emollient; cures dyspepsia, burning sensation, diseases of the blood, biliousness, effects of fatigue, colic, piles, ulcers, menorrhagia. Flowers used in haemorrhagic dysentery.

Satyrium nepalense D. Don

Family: Orchidaceae **Chromosome number(s):** 2n=82

Description: A terrestrial herb attaining a height of 75 cm. Tubers oval, producing stolon each with a small tuber at the end, Stem glabrous, sheathed near the base, Leaves 2-3, mostly basal, fleshy, narrowly elliptic, to 25 cm long and 8 cm broad. Inflorescence many-flowered, to 20 cm long; bracts much longer than the flowers. Flowers 1 cm in diameter, fragrant, pink or white (Bhattacharjee 1998).

Distribution: WCE; alt. 600-4600 m.
Himalayas (Kashmir to Bhutan), India, Sri Lanka, N. Myanmar, W. China.

Conservation status: Threatened (IUCN Category).

Part(s) used: Roots.

Uses: Roots are used to treat malaria and dysentery. It is also used as tonic.

Saurauia napaulensis DC.

Synonym(s): *Saurauia paniculata* Wall. **Common name(s):** Gogan (Nep.)
Family: Saurauiaceae

Description: A large deciduous shrub or small tree with reddish bark and soft spongy wood. Leaves large mostly 18-36 cm, elliptic, with acute gland-tipped marginal teeth and many pairs of prominent lateral veins, young leaves particularly densely rusty-haired. Flowers many, c. 1.3 cm across, in lax axillary clusters, pink, shorter than the leaves; calyx pinkish; petals 5, with fringed margins; styles 4-6, conspicuous. Fruit green, 4-5-lobed, fleshy with edible sweet pulp.

Distribution: WCE; alt. 750-2100 m.
Himalayas (Uttar Pradesh to Arunachal Pradesh), NE India, N. Myanmar, Thailand, Indo-China, W. China.

Part(s) used: Bark.

Uses: Used as poultice to help extraction of splinter embedded in the flesh.

Schima wallichii (DC.) Korth.

Common name(s): Chilaune (Nep.)
Family: Theaceae
Chromosome number(s): 2n=30, 36

Description: A small or large tree with dark gray rugged bark. Leaves leathery evergreen, elliptic-oblong, 10-18 cm, entire or slightly toothed, hairless and reddish vein beneath. Flowers showy, white, fragrant, in terminal clusters of a few, 3-4 cm across, with 5 broadly ovate petals; flower-stalks 1-3 cm; sepals rounded; stamens many; flower buds large, globular. Capsule globular, *c.* 1.5 cm, woody, splitting into 5 valves.

Distribution: CE; alt. 900-2100 m.
Himalayas (Nepal to Bhutan), NE India, W. China.

Part(s) used: Root stock, bark. Leaves and young plants.

Uses: Bark is anthelmintic, rubifacient and irritates skin. Young plant, root stock and leaves are used as antipyretic.

Scutellaria barbata D. Don

Synonym(s): *Scutellaria peregrina* Roxb.
Common name(s): Barbed skullcap
Family: Labiatae
Chromosome number(s): 2n=26

Description: An annual and perennial herb, to 40 cm tall with glabrous, quadrangular creeping stem. Leaves opposite, the lower with short petiole and the upper sessile, blade ovate or lanceolate, obtuse, entire or crenate to 3 cm long and 1.5 cm broad. Flowers axillary in spikes of 7-14 cm long, with entire, ovate or lanceolate bracts, subsessile or spicate. Nutlets very minute, smooth, granulate, or hispidulous.

Distribution: CE; alt. 1300-1500 m. Nepal, India, China, Myanmar.

Part(s) used: Whole plant

Uses: The plant is used in general fatigue, abdominal pain, ascites and pyodermas. It is used externally in snakebites and injuries.

Selinum wallichianum (DC.) Raizada & Saxena

Synonym(s): *Selinum tenuifolium* Wall. ex C. B. Clarke; *S. tenuifolium* var. *filicifolium* (Edgew.) C. B. Clarke
Common name(s): Bhutkesh, Bhazadri (Nep.); Kanthaparna (Sans.); Ragwort (Eng.)
Family: Umbelliferae
Chromosome number(s): 2n=14, 22

Description: A hairless perennial herb to 150 cm tall. Leaves very finely divided into very numerous elliptic, deeply toothed or lobed segments. Lower leaves to 20 cm long, long-stalked, sheathing at the base; the upper smaller and the uppermost reduced to a sheath. Flowers white, in compound hairy umbels, 5-8 cm across; bracts linear or absent; primary rays 15-30, bracteoles 5-10, linear to lanceolate, white-margined, as long as the flowers. Fruit *c.* 4 mm with broad lateral wings and dorsal and intermediate ribs narrowly winged.

Distribution: WCE; alt. 2700-4800 m.
Himalayas (Kashmir, Nepal, Bhutan), NE India, China (Xizang).

Part(s) used: Whole plant.
Uses: Used in cough and cold.

Semecarpus anacardium L.

Common name(s): Bhaláyo (Nep.); Bhalah (Npbh.); Bhallataka (Sans.); Marking nut tree (Eng.)
Family: Anacardiaceae
Chromosome number(s): 2n=60

Description: A tree of 7.5-9 m in height with gray bark and black latex (Adhikari 1999). Flowers greenish yellow in terminal panicles (?). Fluid from the epicarp of the fruit is highly toxic and cause swelling when comes in direct contact with the skin (?).

Distribution: WCE; alt. 150-1200 m.
Himalayas (Sirmore to Sikkim), India, Myanmar, Malaysia, N. Australia.

Part(s) used: Fruit.

Important biochemical constituent(s): An essential oil containing bhilwanol and other compounds has been isolated from the fruit pericarp (Anonymous 1994).

Uses: Fruit is acrid, hot, used as anthelmintic, used in dysentery, fevers, useful in insanity, asthma and acute rheumatism.

Sesamum orientale L.

Synonym(s): *Sesamum indicum* L.
Common name(s): Til (Nep.); Hamo (Npbh.); Tila (Sans.); Sesame (Eng.)
Family: Pedaliaceae
Chromosome number(s): 2n=26, 52

Description: A branched annual herb to 80 cm tall. Leaves simple, variable in shape, upper ones oblong, and toothed and lower ones ovate and pedatisect. Flowers white or pink with darker markings, borne in racemes at the axils of the leaves. Fruit a capsule, oblong, slightly compressed, grooved into 4, to 5 cm long. Seeds black, brown or white, 2.5 to 3mm long and 1.5 mm broad.

Distribution: WCE; alt. 600-2400 m.
Widely cultivated in tropical Africa, Pakistan, India, Nepal, Sri Lanka, E. Asia.

Part(s) used: Seeds.

Important biochemical constituent(s): A fixed oil, carbohydrates and mucilage are isolated from the seeds. Oil contains a crystalline substance sesamin and a phenol compound sesamol (Anonymous 1994).

Uses: Seeds are considered emollient, diuretic, lactagogue and possess properties of a nourishing tonic. They are given in piles, a paste of the seeds mixed with butter is said to be effective in bleeding piles. A decoction of the seed is used as an emmenagogue and is also given in cough. Combined with linseed the decoction of the seeds is used as an aphrodisiac. A paste of seeds is applied to burns, scalds, etc. and a poultice of seeds is applied to ulcers. Powdered seeds are also used in amenorrhea and dysmenorrhea. Oil possesses anti-oxidant properties Seed is edible, also used for cosmetics and manufacture of soap etc.

Shorea robusta Gaertn.

Common name(s): Sal (Nep.); Laptema (Npbh.); Sala (Sans.)
Family: Dipterocarpaceae
Chromosome number(s): 2n=14

Description: A large deciduous tree with smooth dark brown bark and large lax branches. Leaves large leathery, thin, shining pale, ovate-oblong, narrow-pointed, to 30 cm long with 12-15 pairs of strong lateral veins, short-stalked. Flowers small, very fragrant, pale yellow in axillary or terminal white-haired cymose panicles. Fruit ovoid c. 8 mm across, white-haired, with 5 oblong wings, 5-7.5 cm long, which turn brown when, dry.

Distribution: WCE; alt. 150-1500 m.
Subtropical Himalayas (Uttar Pradesh to Assam), India.

Part(s) used: Resin.

Uses: Resin powder is taken three times a day for 20 days in dysentery (Bhattacharjee 1998).

Sida cordifolia L.

Common name(s): Balu (Nep.); Badyanchoh (Npbh); Bala (Sans.); Country mallow (Eng.)
Family: Malvaceae
Chromosome number(s): 2n = 28, 32

Description: An undershrub of about 80 cm - 1.75 m with 2.5 - 6.5 cm long and 2-3.5 cm broad cordate pubescent leaves having dented margin. Calyx bell-shaped, 5-lobed; corolla 5 pale yellow or white, slightly longer than the calyx. Stamens united in a tube separating at the top into short, free anthers bearing filaments. Styles as many as the carpels and carpels without spinous projection.

Distribution: WCE; alt. 500-1100 m. Pantropical.

Part(s) used: Roots and leaves.

Uses: The juice of plants is given in rheumatism, gonorrhoea and spermatorrhea. Leaves are taken as vegetable in bleeding piles. Root juice is used as sedative and cardiac stimulant. It is astringent and cooling tonic (Malla *et al.* 1997). Seeds make general tonic for improving sexual strength. Decoction of roots is administered in fever (Bhattacharjee 1998).

Sida rhombifolia L.

Synonym(s): *Sida compressa* Wall.
Common name(s): Sano chilya (Nep.); Mahabala (Sans.)
Family: Malvaceae
Chromosome numbers: 2n=14, 14+1B, 16, 28, 28+0-7

Description: An erect annual or perennial undershrub, 1.5 m tall. Leaves highly variable, rhomboid-lanceolate to lanceolate, subglabrous above, grey-pubescent or hoary beneath. Flowers yellow or white, axillary, solitary or in pairs, the branch bearing the panicle-like inflorescences. Seeds smooth, black.

Distribution: CE; alt. 100-500 m. Pantropical

Part(s) used: Root, stem and leaves.

Important biochemical constituent(s): Ephedrine in leaves and ephedrine and some other alkaloids in roots.

Uses: The plant is useful in the treatment of tuberculosis and rheumatism. Roots are used in rheumatism and leucorrhoea. Stem is employed as demulcent and emollient. It is used internally as diuretic, febrifuge and also in skin diseases.

Sigesbeckia orientalis L.

Synonym(s): *Sigesbeckia brachiata* Roxb.; *S. glutinosa* Wall.; *S. orientalis* Roxb.
Family: Compositae
Chromosome number(s): 2n=20, 24, 30, 60

Description: An erect, densely pubescent herb to *c.* 1 m tall with opposite, spreading branches. Leaves to 9 cm long and 4.5 cm broad, opposite, wing-petioled, traingular-ovate, acute or obtuse, irregularly toothed, cuneate at base, 3-nerved, densely pubescent on both surfaces, glandular dotted beneath. Heads 0.7 cm in diameter, peduncled in leafy panicles. Flowers yellow; marginal corollas tinged with pink, recurved, 3-toothed at the apex; inner corollas companulate, 5 toothed. Achene curved, black, glabrous, each enclosed in a boat-shaped bract.

Distribution: WCE; alt. 400-2700 m.
Africa, India, Himalayas, Myanmar, China, Malaysia, S. Japan, Oceania, Australia.

Part(s) used: Whole plant.

Uses: The plant is believed to heal gangrenous ulcers and sores. The fresh juice is used to apply over wounds and as it dries it leaves a varnish-like coating. A tincture of the plant is applied externally to cure ringworm and other parasitic infections. It is also considered effective in rheumatism and renal colic. It is a cardiotonic, antiscorbutic and sialagouge. It is also used as anthelmintic.

Smilax aspera L.

Synonym(s): *Smilax capitata* Buch.-Ham. ex D. Don; *S. fulgens* Wall.; *S. maculata* Roxb. ex D. Don
Common name(s): Chopchini (Nep.)
Family: Liliaceae
Chromosome number(s): 2n=32

Description: A tendril-climber with flexuous usually prickly stem. Leaves linear-lanceolate to rounded, glossy, very variable, to 10 cm or more, margin with or without prickly teeth; leaf bases rounded, heart-shaped, or lobed, usually with prickly leaf-stalk; tendril paired, arising from the base of the leaf-stalk. Flowers small, fragrant, white, unisexual, *c.* 5 mm across, in long axillary spike-like clusters of umbels to 15 cm. Fruit a berry, 6-8 mm, blue-black when ripe.

Distribution: WCE; alt. 1200-2600 m.
Widespread from Mediterranean and E. Africa eastwards to India & Sri Lanka.

Part(s) used: Roots.

Uses: Roots of this plant are used as blood purifier and to treat skin diseases (Bhattacharjee 1998).

Solanum anguivi Lam.

Synonym(s): *Solanum indicum* auct.
Common name(s): Bhantaki, Vrihati (Sans.); Poison-berry (Eng.)
Family: Solanaceae
Chromosome number(s): 2n=24

Description: A spiny herb or an undershrub to 8 m tall with much branched, compressed, spiny stem. Leaves large, pinnatifid, ovate, sinuate or lobed. Flowers blue in racemes of cymose clusters. Berries globose, reddish when young or dark yellow, *c.* 8 mm diam; seeds smooth or minutely pitted, *c.* 4 mm diam.

Distribution: WCE; alt. 250-2300 m.
Nepal, India, Malaya, Indo-China, Philippines, Taiwan.
Part(s) used: Whole plant.

Important biochemical constituent(s): Solasonine in leaves and fruits and solanine in roots, leaves and fruits.

Uses: The plant is carminative and also used as expectorant. It has an effect on human-epidermal carcinoma of the naso-pharynx in tissue culture.

Solanum virginianum Dunal

Synonym(s): *Solanum diffusum* Roxb.; *S. surattense* Burm. f.; *S. xanthocarpum* Schrad. & J.C. Windl.
Common name(s): Kantakari (Nep.); Kantakari, Lakshmana (Sans.); Yellow-berried nightshade (Eng.)
Family: Solanaceae
Chromosome number(s): 2n=24
Description: A prickly spreading perennial herb. Leaves with sharp yellow prickles. Flowers purple. Berries yellow when ripe.
Distribution: WCE; alt. 300-900 m.
Himalayas, N. India, China, SE Asia, Malaysia, Australia, Polynesia.
Part(s) used: Fruits and seeds.

Important biochemical constituent(s): Solasonin glycoalkaloid and diosgenin.
Uses: Juice of berries is given in sore throat. Dried fruit is used as carminative. Seeds are administered as expectorant in asthma and cough.

Swertia alata (Royle ex D. Don) C. B. Clarke

Common name(s): Chiraito, Tite (Nep.); Khalu (Npbh.); Chiretta (Eng.)
Family: Gentianaceae
Chromosome number(s): 2n=24, 26

Description: An erect herb to 60 cm tall. Stems 4-angled. Leaves ovate, short-pointed, often sessile. Flowers bright green to yellow with purple veins.

Distribution: WC; alt. 2000-3600 m. Himalayas (Kashmir to Nepal)

Part(s) used: Whole plant.

Uses: The plant is used in fever. Infusion of plant is used as tonic and febrifuge.

Swertia angustifolia Buch.-Ham. ex D. Don var. angustifolia

Synonym(s): *Swertia angustifolia* var. *hamiltoniana* Burkill
Common name(s): Chiraito, Tite (Nep.); Khalu (Npbh.); Ngul tig (Am.); Chiretta (Eng.)
Family: Gentianaceae
Chromosome number(s): 2n=24, 26

Description: An erect herb to 90 cm tall. Leaves lanceolate, narrow at the base, 1-3-nerved. Flowers white or pale blue in panicles Capsule ovate.

Distribution: WCE; alt. 600-2600 m.
Himalaya (Kashmir to Bhutan), N. India, Myanmar, S. China. *S. angustifolia* var. *pulchella* Burkill.: WCE; alt. 2000 m. Himalaya (Uttar Pradesh to Bhutan), India, Myanmar, China. *S. angustifolia* var. *wallichiana* Burkill.: CE; alt. 600 m. Himalaya (Nepal, Sikkim).

Part(s) used: Whole plant.

Uses: The plant is used as blood purifier and febrifuge (Bhattacharjee 1998).

Swertia bimaculata (Siebold & Zucc.) C.B.Clarke

Synonym(s): *Swertia bimaculata* var. *macrocarpa* Nakai; *Ophelia bimaculata* Sieb. Et Zucc.
Common name(s): Chiraito, Tite (Nep.); Khalu (Npbh.); Chiretta (Eng.) **Family:** Gentianaceae
Chromosome number(s): 2n=18, 24, 26

Description: A herb with 4-angled stem. Leaves elliptic- lanceolate, petioled or at least much narrowed at the base 3-nerved. Flowers on long pedicels, white or yellowish green with black spots on.

Distribution: E; alt. 900-2700 m.
Himalayas (Nepal to Bhutan), NE India (Assam, Nagaland), China, Japan.

Part(s) used: Whole plant.

Uses: Substitute for *Swertia chirayita* for its bitter principle.

Swertia chirayita (Roxb. ex Fleming) H. Karst.

Synonym(s): *Swertia chirata* (Wall.) C. B. Clarke; *Gentiana chirayita* Roxb. ex Fleming
Common name(s): Chiraito, Tite (Nep.); Khalu (Npbh.); Tikta, Gya tik (Am.); Kiratatikta (Sans.); Chiretta (Eng.)
Family: Gentianaceae

Chromosome number(s): 2n=20, 24, 26

Description: An erect annual herb, 60-125 cm tall. Stem robust, branching, cylindrical below, 4-angled upwards, containing a large pith. Leaves broadly lanceolate, 5-nerved, subsessile. Flowers greenish yellow, tinged with purple, in large panicles. Capsules egg-shaped, many sided, c. 6 mm diam., sharp pointed. Seeds smooth, many-angled.

Distribution: CE; alt. 1500-2500 m. Himalayas (Kashmir to Bhutan), NE India.

Conservation status: Vulnerable (IUCN Category). **Part(s) used:** Whole plant.

Important biochemical constituent(s): A yellow bitter acid, Ophelic acid ($C_{15}H_{20}O_{13}$) two bitter glucosides, chiratin ($C_{26}H_{48}O_{15}$) and amarogentin ($C_{32}H_{38}O_{16}$) gentiopicrin, phenols and a xanthone, swerchirin ($C_{25}H_{12}O_6$).

Uses: The plant is bitter, stomachic, febrifuge and anthelmintic. It is used to treat diarrhea, malarial fever and weakness etc. (Bhattacharjee 1998). It is an effective tonic and is prescribed in dyspepsia, in the debility of convalescence and generally in cases in which corroborant measures are indicated. It is given as powder, infusion, tincture, or a fluidextract (Anonymous 1976).

Swertia ciliata (D. Don ex G. Don) B. L. Burtt.

Synonym(s): *Swertia purpurascens* (D. Don) C. B. Clarke; *S. purpurascens* Wall.
Common name(s): Chiraito, Tite (Nep.); Khalu (Npbh.); Chiretta (Eng.) **Family:** Gentianaceae

Description: An annual herb to 90 cm tall. Stems teret or 4-lineolate. Leaves oblong or lanceolate, base narrowed. Flowers purple or dark red.

Distribution: WCE; alt. 2800-4000 m. Afghanistan, Himalayas (Kashmir to Sikkim).

Part(s) used: Whole plant.

Uses: Used as substitute for *S. chirayita*.

Swertia multicaulis D. Don

Common name(s): Chiraito, Tite (Nep.); Khalu (Npbh.); Chiretta (Eng.)
Family: Gentianaceae

Description: An annual herb with many short spreading stems, 5-12 cm, arising directly from the stout rootstock. Leaves narrowly spathulate c. 5 cm, narrowed to a long winged leaf-stalk. Flowers slaty-blue, long-stalked, in a much branched inflorescence.

Distribution: CE; alt. 4000-4900 m. Himalayas (Nepal to Bhutan), China (Xizang).

Part(s) used: Whole plant.
Uses: Substitute for *S. chirayita*.

Swertia paniculata Wall.

Synonym(s): *Swertia dilatata* C. B. Clarke; *S. gracilescens* H. Sm.
Common name(s): Chiraito, Tite (Nep.); Khalu (Npbh.); Chiretta (Eng.)
Family: Gentianaceae
Chromosome number(s): 2n=16

Description: An annual herb to 90 cm tall with spreading branches. Leaves oblong or lanceolate. Flowers white, with 2 purple blotches at the base. Capsule long, pointed.

Distribution: WCE; alt. 1500-4000 m.
Himalayas (Kashmir to Bhutan), NE India, Myanmar, China (Xizang).

Part(s) used: Whole plant.
Uses: Substitute for **S. chirayita.**

Symplocos paniculata (Thunb.) Miq.

Synonym(s): *Symplocos chinensis* (Lour.) Druce; *S. crataegoides* Buch.-Ham. ex D. Don; *S. paniculata* Wall.
Common name(s): Lodh (Nep.); Lodhra (Sans.); Sapphire berry (Eng.)

Family: Symplocaceae
Chromosome number(s): 2n=22, 22+1B

Description: A large deciduous shrub or a medium-sizes tree with corky light gray bark and hairy branchlets and leaves. Leaves membranous, with a broad-elliptic to ovate; blade 4-7 cm sharply toothed, impressed with veins above, short stalked. Flowers fragrant, snow-white, borne in cylindrical branched clusters on lateral stems, to 10 mm across, with 5 oblong-elliptic spreading petals. Fruit globular *c.* 5 mm.

Distribution: WCE; alt. 1000-2500 m.
Himalayas (Kashmir to Arunachal Pradesh), Myanmar, Indo-China, China, Korea, Japan.

Part(s) used: Bark.

Uses: Decoction of the bark is used as gargle in bleeding gum. The bark is astringent and is used to control excessive bleeding during menstruation. It also cures digestive disorder, ulcer and eye diseases (Bhattacharjee 1998).

Syzygium cumini (L.) Skeels

Synonym(s): *Syzygium jambolanum* (Lam. DC.
Common name(s): Jamun (Nep.); Gunjhamsi (Npbh.); Jambu (Sans.); Black plum, Indian blackberry , Java plum (Eng.)
Family: Myrtaceae
Chromosome number(s):2n=42-44, 46

Description: An evergreen tree to 30 m tall with a crooked trunk and thick crown. Leaves oval-oblong or oblong-lanceolate, to15 cm long and 7 cm broad, leathery, tough and smooth, upper surface shiny. Flowers to 15 mm across white, sweet-scented, stalkless, arranged in branches of three. Fruits smooth, rounded, shiny berry, purple-black when ripe.

Distribution: WCE; alt. 300-1200 m.
Subtropical Himalayas, India, Sri Lanka, Malaysia, Australia.

Part(s) used: Bark and seed.

Important biochemical constituent(s): Ellagic acid, alkaloid.

Uses: Fresh bark juice mixed with milk is used in diarrhea. Bark is used in sore throat, bronchitis, asthma, ulcer and dysentery. Seed powder (about 15 g) is administered orally thrice a day for three to four months in diabetes (Bhattacharjee 1998).

Tabernaemontana divaricata (L.) R. Br. ex Roem. & Schult.

Synonym(s): *Tabernaemontana coronaria* (Jacg.) Aitom, *Ervatamia divaricate (L.)* Burkill, *Nerium divaricatum* L. and *Nerium coronarium* Jacg.
Common name(s): Tagar (Nep.)
Family: Apocynaceae

Chromosome number(s): 2n=22, 23, 28, 33, 66

Description: A shrub or small tree with quadrangular, glabrous, longitudinally striated stem. Leaves glabrous, short-petioled, opposite, ovate-lanceolate, entire, acuminate. Flowers in terminal cymes, white, corolla tube swollen at the middle.

Distribution: CE; alt. 250-1200 m.
Native of tropical Asia, cultivated throughout the tropics.

Part(s) used:
Alkaloids, triterpenes and steroids.

Uses: Used as incense and in perfumery. Red pulp of the follicles is used for dyeing fabrics.

Tagetes erecta L.

Common name(s): Sayepatri (Nep.); Takswan, Taphoswan (Npbh.); Sthulapushpa (Sans.); African marigold (Eng.)
Family: Compositae
Chromosome number(s): 2n=24, 24, 48

Description: An annual herb attaining about 60 cm in height, erect, branched. Leaves strong scented, pinnately compound. Flowers yellow; rays sometimes 2-lipped or quilled (in some garden flowers).

Distribution: CE; alt. 1800-2000 m. A garden plant of Mexican origin.

Part(s) used: Roots, leaves and florets.
Uses: An infusion of the plant is used against rheumatism, cold and bronchitis. An extract of the roots is used as laxative. The leaves are given in kidney troubles and muscular pains and are applied in boils and carbuncles. The juice of leaves is used in earache. The leaves and florets are used as emmenagogue. Their infusion is prescribed as a vermifuge, diuretic, and carminative. The florets are used in eye diseases and ulcers.

The oil of tagetes has a strong, sweet and lasting odor. The oil from flower is reddish yellow and that from the leaves and stems is greenish yellow with true to nature marigold aroma.

Taraxacum officinale F. H. Wigg.

Synonyms: *Taraxacum officinale* var. *parvulum* Hook. f.
Common name(s): Tukiphool
Family: Compositae
Chromosome number(s): 2n=8, 16, 18, 24, 24+2B, 26, 27, 32, 38, 40

Description: A hardy perennial herb attaining to 30 cm in height. Leaves deeply serrated. Flowers yellow in few-flowered heads terminating in naked, hollow scape.

Distribution: Nepal

Part(s) used: Roots, leaves and flowers.

Uses: Roots are used to increase urine flow, as a laxative and tonic, to treat liver and spleen ailments and to stimulate appetite. Juice of the fresh plants is effective against liver diseases, chronic hepatitis, visceral congestion, intermittent fever and hypochondria. Flowers used as tea are beneficial in heart troubles (Bhattacharjee 1998).

Taxus wallichiana Zucc.

Synonym(s): *Taxus baccata* auct.; *T. baccata* subsp. *wallichiana* (Zucc.) Pilg.; *T. nucifera* auct.
Common name(s): Lodh salla, Talispatra (Nep.); Kande loti (Dolp.); Manduparni, Talisa (Sans.); Yew (Eng.)
Family: Taxaceae
Chromosome number(s): 2n=24

Description: An evergreen, much branched tree, to 30 m tall, but usually around 10 m with spreading branches. Bark thin, scaly, reddish brown. Leaves linear, flat, curved, spine-tipped, mostly spirally arranged, upper surface shining, lower surface pale, mid-rib conspicuous. Male flowers in short stalk, globose heads in the axils of leaves. Female flowers, solitary, axillary, green. Fruit with red fleshy cup or aril nearly covering the seed.

Distribution: WCE; alt. 2300-3400 m.
Afghanistan, Himalayas (Kashmir to Bhutan), NE India, N. Myanmar, Indo-China, W. China, Malaysia.

Dr. Kamal K. Joshi and Prof. Sanu Devi Joshi

Conservation status: Ban for export (under the Forest Act of HMG 1993).
Part(s) used: Bark and leaves.

Important biochemical constituent(s): Taxine alkaloids (Evans 1989), Taxol (a diterpene).

Uses: Taxol is used as anti-tumor agent and also to cure cancer particularly of breast and uterus (Malla *et al.* 1996). Bark and the leaves are bitter, acrid and nauseous. Leaves are very poisonous, but the aril is harmless. It is emmenagogue and antispasmodic. It is also used in asthma and bronchitis. Wood in the Middle Ages was the principle material for making bows (Bhattacharjee 1998).

Terminalia bellirica (Gaertn.) Roxb.

Common name(s): Barro (Nep.); Balah (Npbh.); Bahira, Bibhitaka (Sans.); Bastard myrobalan, Belleric myrobalan (Eng.)
Family: Combretaceae
Chromosome number(s): 2n=24, 26, 48

Description: It is a medium-sized tree attaining a height of about 18-22 m. Bark gray and soft; wood yellow and hard. Leaves 7.5-20 cm long, ovate form cluster at the apex of each branch. Flowers white or pale yellow in terminal spikes. Fruit a drupe up to 2.5 cm in diameter more or less round, clothed with thin hairs.

Distribution: CE, alt. 300-1100 m.
Nepal, India, Sri Lanka, Myanmar, Thailand, Indo-China, Malaysia.

Part(s) used: Fruits.
Important biochemical constituent(s): ß-sitosterol, gallic acid, ellagic acid, ethyl gallate, galloyl glucose, chebulagic acid, manitol, galactose, fructose, rhamnose and 13.0% tannin.

Uses: Tonic, astringent, laxative, antipyretic, narcotic. Used in piles and dropsy. Fruits are used in cough, hoarseness and eye disease. It is one of the constituents of "Triphala" of Ayurvedic preparation used for liver and gastrointestinal tracts.

Terminalia chebula Retz.

Synonym(s): *Myrobalanus chebula* Gaertn.
Common name(s): Harro (Nep.); Halah (Npbh.); Haritaki, Harra (Sans.); Chebulic myrobalan (Eng.)
Family: Combretaceae
Chromosome number(s): 2n=14, 24, 26, 48, 72

Description: A medium-sized deciduous tree. Leaves ovate, acute, to 20 cm long, with two small glands near the leaf base. Flowers in terminal spike at the branches, dull white. Fruit 5-ribbed, to 4 cm long.

Distribution: CE; alt. 150-1100 m.
Himalayas (Uttar Pradesh, Nepal), India, Sri Lanka, Myanmar.

Part(s) used: Fruits.

Important biochemical constituent(s): Large quantity of tannins. Also contains chebulagic acid, chebulinic acid and corilagin.

Uses: Fruit is astringent and laxative. It is used as gargle in the inflammation of mucous membrane of the mouth. Decoction of fruits is used in bleeding and ulceration of gums. Fruit is roasted and eaten three times a day for one week in cough. It is also used externally to treat chronic ulcer, wounds and scalds (Bhattacharjee 1998).

Tinospora sinensis (Lour.) Merr.

Synonym(s): *Tinospora cordifolia* auct.; *T. malabarica* (Lam.) Hook. f. & Thomson; *T. tomentosa* (Colebr.)Hook. f. & Thomson
Common name(s): Gurjo (Nep.); Bhote khair, Umma lauri, Kuchrung (Dolp); Guduchi (Sans.); Gulancha tinospora (Eng.)
Family: Menispermaceae
Chromosome number(s): 2n=24, 26

Description: A monoecious dichlinous climber with succulent stem and aerial roots. Stem and branches with white glands. Leaves ovate, or nearly round, to 10 cm long. Flowers very small; female flowers solitary and the male flowers in groups at the axils of bracts.

Dr. Kamal K. Joshi and Prof. Sanu Devi Joshi

Distribution: EC; alt. 300-500 m.
Nepal, NE India, S. India, Sri Lanka, Myanmar, Thailand, Vietnam, S. China, Malaya.

Part(s) used: Roots, bark and stem.

Important biochemical constituent(s): Giloin, gilenin, gilosterol and tinosporine have been isolated from the stem (Anonymous 1994).

Uses: Starch of roots and stem is nutritious and are used to cure diarrhea. Dried bark and the stem are used as tonic. It is antipyretic and aphrodisiac.

Trachyspermum ammi (L.) Sprague

Common name(s): Jwano (Nep.); Imoo (Npbh.); Yavani (Sans.); Ajowan (Eng.)
Family: Umbelliferae
Chromosome number(s): 2n=18

Description: An erect, glabrous or minutely pubescent, branched annual herb, to 90 cm tall with striate stems. Leaves 2-3-pinnately divided with linear segments. Flowers white, small in terminal or seemingly lateral pedunculate, compound umbels. Fruits ovoid, muricate, aromatic; cremocarps, 2-3 mm long, grayish brown; mericarps compressed, with distinct ridges and tubercular surface, 1-seeded.

Distribution: C; alt. 1000 m.
Nepal. Very widely cultivated in India for medicinal purposes.

Part(s) used: Dried fruits.
Important biochemical constituent(s): Fruits have aromatic essential oil containing cumene, thymol and thymene.

Uses: Fruits are antispasmodic, stomachic, carminative, stimulant and tonic. Used in diarrhea, dyspepsia, colic, flattulence, indigestion and cholera.

Tribulus terrestris L.

Common name(s): Gokhur (Nep.); Gokshura (Sans.); Calthrops (Eng.) **Family:** Zygophyllaceae
Chromosome number(s): 2n=12, 24, 32, 36, 48

Description: A trailing and spreading herb, densely covered with minute hairs. Leaves compound, opposite, leaflets 3-6 pairs, to 8 cm long. Flowers usually silky, white or yellow, solitary, arising from the axil of the leaf. Ovary bristly; style short and stout. Fruits globose, spinous or tuberculate, hairy or nearly glabrous, often muriculate, woodi cocci, each with two pairs of hard sharp spines, often cling to clothes and in the fur of animal bodies. Seeds many in each of woodi cocci.

Distribution: C; alt. 150 m. Pantropical.

Part(s) used: Roots, leaves and fruits.

Important biochemical constituent(s): Steroidal saponin, diosgenin, -sitosterol and stigmasterol were isolated from this plant.

Uses: The drug is diuretic, tonic and aphrodisiac. Roots are stomachic, appetizer, diuretic and carminative. Decoction of leaves is used as gargle in painful gums and inflammation in mouth. Leaves increase the menstrual flow and cure gonorrhoea. Fruits are useful in urinary complaints, painful micturition and impotence. Fruits are also used to treat cough, and scabies (Bhattacharjee 1998).

Trichosanthes tricuspidata Lour.

Synonym(s): *Trichosanthes bracteata* (Lam.) Voigt; *T. lepiniana* (Naudin) Cogn.; *T. palmata* Roxb.; *T. pubera* Blume; *T. quinquangulata* A. Gray
Common name(s): Indrayani (Nep.); Shwetpushpi (Sans.)
Family: Cucurbitaceae

Description: A herbaceous climber with branched tendril. Leaves heart-shaped, to 20 cm, deeply 3-7-lobed. Flowers white, to 8 cm across, with petals and bracts very conspicuously long-fringed. Fruit globular, to 5 cm, red striped with orange; seeds in dark green pulp.

Distribution: WCE; alt. 1200-2300 m.

Himalayas, India, east to China, Japan, Malaysia, tropical Australia.

Part(s) used: Fruits and seeds.

Important biochemical constituent(s): An amorphous saponin, trichosanthin, hentriacontane, essential oil containing terpenes etc.

Uses: It is used as carminative, purgative, and abortifacient. It is also used in asthma and earache. It lessens inflammation; cures hemicrania, weakness of limbs and heat of brains. The seeds are emetic and purgative.

Typha angustifolia L.

Synonym(s): *Typha elephantina* Roxb.; *T. javanica* Schnizi. ex Rohrb. **Common name(s):** Elephant-grass (Eng.)
Family: Typhaceae
Chromosome number(s): 2n=30

Description: An aquatic gigantic, gregarious, marshy plant, to *c.* 3.5 m tall. Rhizomes thick, stoloniferous. Leaves grass-like, spongy, trigonous, deep green, very smooth, to 1.8 m long and 3.8 cm broad. Male spikes longer than female; anther yellow with green tops; pollen 4-globate; female flowers on stouter spikes, brown, to 25 cm long and 2.5 cm broad, mixed with clavate pistillodes, perianth hairs fine, capillary. Seeds oblong.

Distribution: E; alt. 200 m.
Temperate and subtropical regions in Europe, Asia and N. America.

Part(s) used: Rhizomes, pollen, male spikes and ripe fruits.

Uses: Rhizomes are somewhat astringent and diuretic and are given in dysentery, gonorrhoea and measles. The pollen is substituted for the spores of *Lycopodium* spp. The soft and woolly floss of male spikes and the down of the ripe fruits were used in emergency as medicated absorbent to wounds and ulcers.

Uraria picta (Jacq.) Desv. ex DC.

Common name(s): Prishniparni (Nep.) Prishniparni, Chitraparni (Sans.)
Family: Leguminosae
Chromosome number(s): 2n=16, 22

Description: An erect perennial herb to 180 cm tall. Leaves 1-3 foliate, to 30 cm long, blotched white; leaflets 4-6, rarely 9, linear-oblong or lanceolate, obtuse, mucronate, stipules subulate. Flowers purple, in dense cylindrical racemes. Pods 3-6 -jointed, polished, often whitish, glabrous.

Distribution: W; alt. 1000-1100 m.
Tropical Africa, Himalayas, India, Sri Lanka, SE Asia, China, Malaysia, Australia.

Part(s) used: Roots, leaves and pods.

Uses: One of the components of Dashmularishta and Chyawanprashavaleha, the Ayurvedic preparations used for body pains including the rheumatic pains, and fever. The plant is credited with fracture-healing properties. The root is said to aphrodisiac. Its decoction is given in cough, chills and fevers. The leaves are considered antiseptic and used in gonorrhoea. Roots and pods are used to treat prolapse of anus in infants and sore-mouth in children.

Urtica dioica L.

Common name(s): Sisnu (Nep.); Nhayekan (Npbh.); Stinging-nettle (Eng.)
Family: Urticaceae
Chromosome number(s): 2n=32, 48, 52

Description: A robust dioecious herb, to 2 m tall, with grooved stems, abundantly armed with stinging hairs. Leaves ovate or lanceolate, usually cordate, serrate. Flowers greenish in axillary cymes.

Distribution: WC; alt. 3000-4500 m.
Europe, N. America, W. Siberia, C. Asia, Himalayas, W. China, naturalized widely in other temperate regions.

Part(s) used: Whole plant.

Uses: The plant is haemostatic in vomiting of blood, uterine haemorrhage and bleeding from the nose. It is also used in sciatica, palsy and rheumatism. Drops of root extract are applied in hollow cavity of tooth during toothache. It is also given for easy delivery. Paste of root is astringent (Bhattacharjee 1998).

Usnea thomsonii Stirt.

Common name(s): Jhyau, Charila, Budhna (Nep.); Lichen (Eng.)
Family: Usneaceae

Description: Thallus fruticose with central solid axis, 10 cm tall, non-pigmented or yellowish-brown to brownish-black, erect or procumbent, pseudocyphellate, some times articulate and cracked. Lacks soredia and isidia. Medulla K-, Apothecia common. Spores to 19 mm long and 12 mm wide.

Distribution: WCE; alt. *c.* 3000 m Himalayas Nepal, India.

Conservation status: Under the Forest Act 1993, His Majesty's Government of Nepal banned on export of Lichen in crude form without processing. However, processed Lichen resinoid is allowed to export.

Part(s) used: Whole plant.

Uses: Used in mental ailments including epilepsy. It is also used in incense sticks, spices and veterinary drugs. The paste is used as ointment and antibiotic in cuts and wound. Lichen resinoid used as a fixative in high-grade perfume.

Valeriana jatamansii Jones

Synonym(s): *Valeriana spica* Vahl; *V. villosa* Wall.; *V. wallichii* DC. **Common name(s):** Sugandhawal (Nep.); Indian valerian (Eng.) **Family:** Valerianaceae
Chromosome number(s): $2n=28$

Description: A perennial herb to 45 cm tall with, thick, horizontal, nodular and aromatic rootstock and tufted stem. Flowers white with pink tinge.

Distribution: WCE; alt. 1500-3300 m.
Afghanistan, Himalayas (Kashmir to Bhutan), NE India, Myanmar, W. and C. China.

Conservation status: Ban for export (under the Forest Act of HMG 1993) **Part(s) used:** Roots and leaves.

Uses: Roots are used in afflictions of eyes and blood and enlargement of liver and spleen. Root preparations are used in cosmetics and hair oil. Crushed leaves are rubbed on forehead during severe headache.

Vanda cristata Lindl.

Synonym(s): *Vanda alpina* auct.
Family: Orchidaceae
Chromosome number(s): n=19

Description: An epiphytic herb of 15 cm tall with erect and stout stem. Leaves narrowly oblong, bilobed at the apex, to 15 cm long and 1.5 cm broad, horizontally arranged and leathery. Inflorescence erect, shorter than leaves, 3-6-flowered. Flowers 5 cm across, yellowish or olive green with purplish blotches (Bhattacharjee 1998).

Distribution: CE; alt. 1200-2300 m.
Himalayas (Uttar Pradesh to Bhutan), NE India (Meghalaya).

Conservation status: Threatened (IUCN Category).
Part(s) used: Leaves.

Uses: Leaves are used as expectorant.

Viscum album L.

Synonym(s): *Viscum costatum* Gamble; *V. stellatum* Buch.-Ham. ex D. Don

Common name(s): Harchur, Ainjeru (Nep.); Harchu (Npbh.); Mistletoe, Devil's fuge (Eng.)
Family: Loranthaceae
Chromosome number(s): 2n=20

Description: A yellowish green, dichotomously branched, semiparasitic, evergreen, dioecious shrub, to 120 cm tall. Leaves to 10 cm long, lanceolate or elliptic, entire. Flowers in terminal fascicles. Pseudo-berry white, elipsoid, viscid, with 1 seed embedded in the pulp.

Distribution: WC; alt. 600-2300 m. Himalayas (Kashmir to Nepal), India.

Part(s) used: Whole plant.

Uses: All parts of plant are diuretic.

Viscum articulatum Burm. f. var. articulatum

Synonym(s): *Viscum nepalense* Spreng.
Common name(s): Harchur (Nep.); Harchu (Npbh.); Gandhmadini (Sans.); Mistletoe (Eng.)
Family: Loranthaceae
Chromosome number(s): 2n=24

Description: A much-branched, pendulous, leafless, parasitic, green shrub. Host chiefly Acacia ferruginea, Dalbergia latifolia, Diospyros ebenum, D. melanoxylon, Grewia rotundifolia and Ixora arborea. Stems flattened, jointed. Flowers green, male and female in the same cluster. Fruit yelow, smooth, juicy, with a bright green seed.

Distribution: WC; alt. 1500-2100 m.
Himalayas (Nepal to Bhutan), India, Myanmar, Sri Lanka, China, Indo-China, Malaysia. *V. articulatum* var. *liquidambaricolum:* WCE; alt. 200-2200 m. Himalayas (Himachal Pradesh to Arunachal Pradesh), India, Myanmar, Indo-China, China.

Part(s) used: Whole plant.

Important biochemical constituent(s): Oleanolic acid, ceryl pleanolate and *meso*-inositol.

Uses: The plant is believed to have aphrodisiac and febrifuge properties. Its paste is applied to cuts.

Vitex negundo L. var. negundo

Common name(s): Simali (Nep.); Nirgundi (Sans.); Indian privet (Eng.)
Family: Verbenaceae
Chromosome number(s): 2n=24, 26, 28, 32, 34

Description: A large aromatic shrub with quadrangular densely whitish tomentose branches, to 4.5 m tall. Bark thin, gray. Leaves 3-5 foliate; leaflets lanceolate, entire or rarely crenate, terminal leaflets to 10 cm long and 3.2 cm broad, lateral leaflets smaller, all nearly glabrous above, white tomentose beneath. Flowers bluish purple, small, in peduncled cymes, forming large, terminal, often compound, pyramidal panicles. Fruit a drupe, globose, black when ripe, 5-6 mm diam. with persistent calyx at the base.

Distribution: WCE; alt. 100-1200 m.
Himalayas (Nepal to Bhutan), Afghanistan, India, Sri Lanka, China, Myanmar, Indo-China, Malaysia.

Part(s) used: Leaves.

Uses: Leaves are aromatic and are considered tonic and vermifuge. A decoction of the leaves, with the addition of long pepper *(Piper longum)* is given in catarrhal fever with heaviness of the head and dullness of hearing. The leaves are also smoked for the relief from headache and catarrh. An ointment made from the juice of leaves is applied as a hair-tonic.

Woodfordia fructicosa (L.) Kurz.

Synonym(s): *Woodfordia floribunda* Salisb.
Common name(s): Dhainyaro, Amar phool (Nep.); Dhataki (Sans.); Fire-flame bush (Eng.)
Family: Lythraceae
Chromosome number(s): 2n=16

Description: A much-branched, beautiful shrub, with fluted stems and long, spreading branches, to 3 m tall, rarely to 7 m. Bark reddish brown, peeling off in thin, fibrous strips. Leaves lanceolate, oblong-lanceolate or ovate-lanceolate. Flowers numerous, brilliant red in dense axillary paniculate-cymose clusters. Fruit a capsule, ellipsoid, membranous; seeds brown, minute, smooth, obovate.

Distribution: WCE; alt. 200-1800 m.
Africa, W. Asia, subtropical Himalayas, India, Sri Lanka, Myanmar, east to China.

Part(s) used: Leaves and flowers.

Important biochemical constituent(s): Leaves and flowers contain elegiac acid and polyphenols.

Uses: Leaf paste is used in skin diseases and leaf juice is applied in conjunctivitis. Dried flowers are astringent and used in dysentery. They are also used in affliction of mucous membrane of bilious complaints.

Wrightia arborea (Dennst.) Mabb.

Synonym(s): *Wrightia mollissima* Wall.; *Wrightia tomentosa* Roem. & Schult.
Common name(s): Khirro, Karingi (Nep.)
Family: Apocynaceae
Chromosome number(s): 2n=22

Description: A small tree attaining to 9 m tall with yellow milky juice, opposite divaricate scabrous branches, young parts of the bark densely tomentose. Leaves to 12 cm long and 6.5 cm broad, elliptic-oblong, acuminate, tomentose. Flowers 2.5 cm or more across, in erect, corymbose, tomentose cymes, white, turns yellow shortly after being picked up. Fruit cylindric, to 30 cm long. Seeds to 1.5 cm long, slender, attenuated at the apex.

Distribution: CE; alt. 450-950 m.
Himalayas, India, Sri Lanka, Myanmar, Thailand, Indo-China, S. China.

Part(s) used: Bark.

Uses: Decoction of bark is given in menstrual and renal complaints.

Xeromphis spinosa (Thunb.) Keay

Synonym(s): *Randia spinosa* (Thunb.) Poir.
Common name(s): Mainphal (Nep.); Madanam (Sans.); Common emetic nut (Eng.)
Family: Rubiaceae
Chromosome number(s): 2n=22

Description: A large spiny deciduous shrub or small tree. Leaves obovate, to 5 cm long; spines to 4 cm, axillary. Flowers solitary, fragrant, white, borne on short leafy branchlets. Fruits to 4 cm across, globose or ovoid, often crowned with persistent calyx.

Distribution: W; alt. 100-1200 m.
Himalayas, India, Indo-China, S. China, Malaysia.

Part(s) used: Bark, stem and fruits.

Uses: Paste of the stem is applied on joint-ache. Bark is used to treat diarrhea, dysentery, fever and rheumatism. The fruit pulp is anthelmintic and abortifacient and the dried fruit powder is used as emetic while the ripe fruit is used as fish poison.

Zanthoxylum armatum DC.

Synonym(s): ***Zanthoxylum alatum*** Roxb.; ***Z. hostile*** Wall.; ***Z. violaceum*** Wall.; ***Z. oxyphyllum*** Edgew.
Common name(s): Timur (Nep.); Timoo (Npbh.); Tumburu, Dhiva, Gandhalu (Sans.)

Family: Rutaceae
Chromosome number(s): 2n=66

Description: A shrub or rarely small tree, with corky bark and numerous long straight spines on aromatic branchlets and narrowly winged leaf-stalks. Leaves pinnate; leaflets 2-6 pairs, lanceolate, *c.* 8 cm long, toothed, sparsely gland dotted. Flowers *c.* 1 mm, one-sexed; calyx with 6-8 acute lobes; petals absent; stamens 6-8, much longer than calyx in male flowers. Ripe capsules 3-4 mm diam., globular, red, wrinkled, aromatic; seed shining black.

Distribution: WCE; alt. 1100-2500 m.
Himalayas (Kashmir to Bhutan), N. India, east to China, Taiwan, Philippines, Lesser Sunda Islands.

Part(s) used: Stem, leaves and fruits.

Important biochemical constituent(s): Tannic acid and gallic acid, starch, mineral salts, mucilage and albumen have been isolated from the rhizome (Anonymous 1994).

Uses: Stem is used as tooth brush and mouth purifier. Leaves and fruits are chewed in teeth enamel disease (Bhattacharjee 1998).

Ziziphus mauritiana Lam.

Synonym(s): *Ziziphus jujuba* (Lam.) Gaertn.
Common name(s): Bayar (Nep.); Bayeli (Npbh.); Ajapriya, Badara, Karkandhu, Kuvala, Madhraphala (Sans.); Common jujube, Chinese date (Eng.)
Family: Rhamnaceae
Chromosome number(s): 2n=48

Description: A shrub to 4 m tall with spreading and drooping tomentose branches armed with curved prickles. Leaves ovate or orbicular, to 7 cm long and 4.5 cm broad, prominently 3-veined, upper surface green, glabrous, lower gray, tomentose, entire or minutely serrate, unequal at the base. Flowers greenish or yellowish, c. 3 mm across, in axillary clusters. Petals spathulate, concave, reflexed. Fruit a drupe, globose, red when ripe, each with a single stone surrounded by fleshy pulp.

Distribution: WCE; alt. 200-1200 m. Tropical Asia, Australia; widely cultivated.

Part(s) used: Barks, leaves and fruits.

Uses: Root is used to treat biliousness. Bark is applied in boils and is used internally in dysentery and diarrhea. Leaves are antipyretic and reduce obesity. Fruit is cooling, aphrodisiac, tonic, laxative and useful in blood diseases.

Chandanbari village with *Abies spectabilis* forest

Abies spectabilis with young fruits

Aconogonum rumicifolium sp. (Mount Annapoorna in the background)

Acorus calamus

Achyranthes aspera

Aconitum spicatum

Medicinal and Aromatic Plants Used in Nepal, Tibet and Trans-Himalayan Region

Alnus nepalensis in Suhki Pokhari, Dhankuta

Alnus nepalensis (canopy)

Amaranthus spinosus

Amomum subulatum under the shade of *Alnus nepalensis*

Arisaema tortuosum

Arisaema tortuosum (fruit)

Medicinal and Aromatic Plants Used in Nepal, Tibet and Trans-Himalayan Region

Artemisia indica

Asparagus recemosus

Astilbe rivularis

Astilbe rivularis (rhizome)

Bauhinia variegata (flowers)

Medicinal and Aromatic Plants Used in Nepal, Tibet and Trans-Himalayan Region

Berberis asiatica

Bergenia ciliata

Betula utilis forest

Medicinal and Aromatic Plants Used in Nepal, Tibet and Trans-Himalayan Region

Betula utilis (close-up of trunk)

Butea buteiformis

Butea buteiformis (an inflorescence)

Cannabis sativa

Cassia fistula

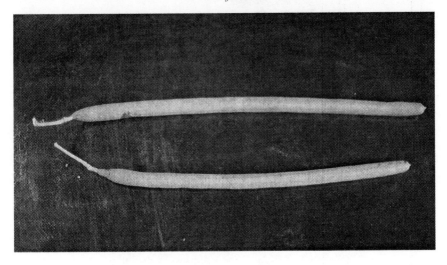

Pods of *Cassia fistula*

Cinnamomum glaucescens

Cinnamomum tamala

Dactylorhiza hatagirea

Dactylorhiza hatagirea

Dr. Kamal K. Joshi and Prof. Sanu Devi Joshi

Daphne bholua

Datura metel

Datura stramonium

Datura stramonium with fruits

Dioscorea bulbifera

Elaeocarpus sphaericus

Elaeocarpus sphaericus (the flowers)

Elaerocarpus sphaericus (the fruits)

Medicinal and Aromatic Plants Used in Nepal, Tibet and Trans-Himalayan Region

Ephedra girardiana

Fritillaria cirrhosa

Geranium nepalense

Hedera nepalensis

Heracleum napalense

Hippophae tebatana

Juglans regia (nuts)

Juniperus indica (powdered stem)

Justicia adhatoda

Lobelia pyramidalis

Mahonia napaulensis

Mangifera indica

Murraya koenigii

Ocimum tenuiflorum

Paris polyphylla

Paris polyphylla

Medicinal and Aromatic Plants Used in Nepal, Tibet and Trans-Himalayan Region

Parnassia nubicola

Phyllanthus emblica (fruits)

Podophyllum hexandrum

Potentilla microphylla

Punica granatum (fruit)

Rauvolfia serpentina

Rheum australe

Rhododendron anthopogan

Medicinal and Aromatic Plants Used in Nepal, Tibet and Trans-Himalayan Region

Rhododendron arboreum

Rhododendron lepidotum

Rosa laevigata

Riccinus communis

Selenium wallichianum

Sesamum orientale

Shorea robusta

Dr. Kamal K. Joshi and Prof. Sanu Devi Joshi

Swertia chirayita

Syzygium cumini

Taxus wallichiana

Taxus wallichiana (a fruiting branch)

Trichosanthus sp.

Medicinal and Aromatic Plants Used in Nepal, Tibet and Trans-Himalayan Region

Trichosanthus sp.

Urtica dioica

Valeriana jatamansii

Zanthoxylum armatum

Zanthoxylum armatum (with fruits)

GLOSSARY OF MEDICAL TERMS

Abortifacient: a drug causing abortion or miscarriage.

Abscess: collection of pus, a thick yellowish-white fluid caused by bacterial, protozoan or fungal invasion of body tissues.

Acrid: unpleasantly pungent or sharp to smell or taste.

Alexipharmic: acting as an antidote.

Alterative: stimulating change in metabolism and tissue function in defence against diseases, not chronic and acute.

Amenorrhea: amenia; absence of abnormal cessation of menses.
Amnesia: partial or total loss of memory.

Anaemia: a blood condition involving an abnormal reduction in the number of red blood cells (erythrocytes) or in their haemoglobin content.

Analgesic: a pain reliever that does not induce loss of consciousness; anodyne.

Ancylostomiasis: hookworm disease; Egyptian or tropical chlorosis; tunnel disease or anaemia; miner's anaemia; brickmaker's anaemia; pronounced anaemia, emaciation, dyspepsia, and swelling of the abdomen with mental and the physical inertia due to the presence of a species of *Ancyclostoma* or *Necator* in the intestine.

Anesthesia: induction of the loss of tactile sensibility (numbness), especially of pain.

Angioneurotic oedema: periodically recurring episodes of noninflammatory swelling of skin, mucous membranes, viscera and brain of sudden onset and lasting hours to days, occasionally with arthralgia (sever pain in a joint), purpura (a condition characterized by haemorrhage into the skin) or fever.

Anodyne: analgesic.

Anthelmintic: a drug used for combating intestinal worms.

Antihistamine: a drug that blocks the action of histamine. Antihistamines are used primarily to control symptoms of allergic conditions such as hay fever, runny nose, sneezing, conjunctivitis and breathing difficulties.

Antioxidant: an agent that inhibits oxidation and thus prevents rancidity (the state of having disagreeable odor or taste) of oils or fats or the deterioration of other materials through oxidative processes.

Antipyretic: a drug that relieves or reduces fever.

Antiseptic: an agent that destroys or checks germs.

Antispasmodic: a drug that relieves muscular spasm and has a sedative effect on the nerves.

Antitussive: an agent that prevents or stops convulsion (muscular relaxant).

Aperient: mild laxative.

Aphonia: loss of voice in consequence of disease or injury of the organ of speech.

Aphrodisiac: a drug that stimulates sexual desire.

Aphtha: a minute ulcer on a mucous membrane of the mouth, often covered by a gray or white exudate.

Aromatic: a fragrant or spicy or mildly stimulant drug.

Arthralgia: arthrodynia; severe pain in a joint, especially one not inflammatory in character.

Ascariasis: disease caused by infection with *Ascaris* or related ascarid nematodes.

Ascites: accumulation of serous fluid in the peritoneal cavity.

Asthenia: weakness, debility.

Asthma: a chronic inflammation of the bronchial tubes.

Astringent: an agent that induces shrinking and hardening of tissues, thereby lessening secretion.

Attar: Rose oil from **Rosa** spp.

Biliousness: a disturbance in the digestive system due to improper functioning of the liver.

Bronchitis: inflammation of the membrane lining of the bronchial tube. **Bronchodilator:** an agent that causes an increase in caliber of a bronchus or bronchial tube.

Cancer: any malignant growth of tissues caused by uncontrolled proliferation of corresponding cells.

Carbuncles: localized inflammation of tissue under the skin caused due to infection usually by the bacteria **Staphylococcus.** The infected part is larger and more serious than a boil, which results from an inflammation beginning in an infected hair sac.

Carminative: an agent that promotes expulsion of gas (or reduces the formation of gas) in the stomach.

Catarrh: the discharge from inflamed mucous membranes, such as those lining the nasal cavity or stomach.

Cathartic: purgative, evacuant; a fairly powerful purgative (an agent used to evacuate the bowel).

Cephalagia: headache.

Chyavanprasha: an Ayurvedic preparation of many herbs, the chief of them being the fruits of **Phyllanthus emblica** as an important source of vitamin C. Chyavanprasha is specifically indicated in the cases of bronchial asthma, tuberculosis, scurvy and nasal polyp (i.e., polypus; a general descriptive term used with reference to any mass of tissue that bulges or projects outward, or upward, from the normal surface level, thereby being macroscopically visible as a hemispheroidal, spheroidal or irregularly mound-like structure growing from a relatively broad base or a slender stock).

Cholagogue: an agent that promotes the flow of bile into the intestine, especially as a result of contraction of the gallbladder.

Cholera: a disease marked by diarrhea and loss of water with characteristic "rice-water stools," vomiting, thirst, muscle cramps, and sometimes circulatory collapse.

Cirrhosis: chronic (and often fatal) disease of liver.

Colic: pain due to spasmodic contraction of the abdomen.

Colitis: an inflammatory disease of the colon.

Conception: conceiving; becoming pregnant.

Conjunctivitis: inflammation of the conjunctiva (a mucous membrane that lines the inner surface of the eyelids and joins with the cornea of the eyeball).

Constipation: infrequent or difficult bowel movements characterized by dry, hardened feces.

Convalescence: gradual recovery of health and strength after an illness.

Convulsion: series of involuntary contractions of the voluntary muscles. The eyeballs frequently roll upward or to one side during a convulsion; breathing appears labored, and saliva oozes from the mouth. The teeth are usually tightly clenched, sometimes causing serious bites to the tongue and the cheeks.

Corneal ulcer: ulcer (open sore) on cornea (a tough transparent outer membrane protecting the iris of the eyeball).

Cystitis: inflammation of the urinary bladder, usually from the bacterial infection originating in the urethra, vagina, or in more complicated cases, the kidneys. It may also be caused by irritation from crystalline deposits in the urine or from any condition or urologic abnormality that interferes with normal bladder function.

Dashang: a paste prepared from the bark of **Albizia lebbeck** used in Ayurvedic medicine.

Dasmula: an Ayurvedic preparation comprising a group of roots 10 plants.

Decoction: an extract of a crude drug obtained by boiling in water. **Delirium:** an extreme mental disturbance marked by excitement, restlessness and rapid succession of confused and unconnected ideas.

Demulcent: a drug having a soothing action on the skin and mucous membrane.

Deobstruent: an agent that removes an obstruction to secretion or excretion.

Depurative: tending to depurate; depurant (an agent or means used to effect purification; an agent that promotes the excretion and removal of waste material.

Diabetes: a metabolic disorder affecting the production of insulin resulting in the excess flow of sugar in the urine and consequent rise in the level of glucose in blood.

Diarrhea: abnormal frequency and fluidity of faecal discharges. **Diphtheria:** an infectious disease of the throat and the air passage. **Diuretic:** a drug that increases the secretion and discharge of urine. **Dropsy:** a disease marked by an excessive collection of a watery fluid in body tissues or cavities.

Dysentery: an infectious disease causing acute diarrhea and discharge of mucus and blood.

Dyspepsia: the impaired digestion of food.

Dyspnea: shortness of breath; subjective difficulty or distress in breathing, frequently rapid breathing, usually associated with serious disease of the heart or lungs.

Dystocia: difficult childbirth.

Eczema: an inflammatory, chronic, noncontagious disease of the skin caused by allergy and hypersensitivity.

Electuary: a medicine that melts in the mouth.

Emetic: an agent that induces vomiting.

Emmenagogue: a drug that promotes menstruation or regulates the menstrual period.

Emollient: a drug that allays irritation of the skin and alleviates swelling and pain.

Emphysema: progressive respiratory disease characterized by coughing, shortness of breath, and wheezing, developing into extreme difficulty in breathing, and sometimes resulting in disability and death.

Epilepsy: a chronic nervous disorder marked by attacks of unconsciousness or convulsions.

Epistaxis: an ailment characterized by the bleeding from the nose usually associated with nasal polyp.

Expectorant: a drug that helps the removal of catarrhal matter and phlegm from the bronchial tubes.

Febrifuge: antipyretic; an agent that reduces fever.

Flatulence: a disorder causing excessive accumulation of gas in the alimentary canal.

Gastralgia: stomachache.

Gastrorrhagia: haemorrhage from the stomach.

Glaucoma: a family of diseases, characterized by abnormal pressure within the eyeball leading to the loss of visual field and declining vision.

Goitre: a chronic enlargement of the thyroid gland.

Gonorrhoea: an infectious veneral disease marked by a burning sensation when urinating and a mucopurulent discharge from the urethra or vagina.

Gout: painful disease affecting joints due to faulty purine metabolism.
Haematemesis: vomiting of blood.

Haematuria: presence of blood in the urine commonly caused by stones or infection in the genito-urinary tract or some other hemorrhagic conditions.

Haemoptysis: splitting of blood while coughing primarily caused by such diseases as tuberculosis and cancer of the lungs.

Haemorrhage: profuse bleeding from ruptured blood vessel.

Hepatitis: inflammation of the liver, usually from a viral infection, sometimes from toxic agents.

Hoarseness: inflammation of the throat including pharynx and the larynx causing roughness and harshness of the voice.

Hypertension: abnormally high constructive tension in blood vessels; usually revealed as high blood pressure.

Hypochondrium: a traditional term for a morbid condition characterized by the stimulation of the symptoms of any of several diseases.

Hypotension: a state of abnormally low blood pressure.

Hysteria: an affliction of losing one's control over acts and feelings and suffering from imaginary ailments.

Icteric: relating to or marked by icterus (jaundice).

Impotence: state of being impotent (lack of sufficient strength or unable to act of males; wholly lacking in sexual power.

Infusion: an extract obtained by steeping the drug in water.

Insanity: madness.

Insomnia: condition in which a person has difficulty getting sufficient sleep.

Intoxicant: a drug that leads to intoxication (state of losing self-control as the result of taking alcohol or other agent).

Jaundice: a condition characterized by a raised level of bilirubin (breakdown product of haemoglobin), producing yellowness of mucous membranes including the eyes.

Keratitis: inflammation of the cornea.

Laryngitis: a disease causing inflammation of the larynx.

Laxative: a mild purgative.

Leprosy: an infectious disease characterized by ulcers and white, scaly scabs.

Leucoderma: any area of the skin that is white, congenital or acquired absence or loss of melanin pigmentation.

Leucorrhoea: whitish, purulent mucous discharge from the vagina and uterine canal.

Lithiasis: the so-called uric acid diathesis (arrangement, condition; the constitutional or inborn state disposing to a disease, group of diseases, or metabolic or structural anomaly); the formation of calculi of any kind, especially of biliary or urinary calculi.

Lumbago: pain in lower back.

Migraine: periodic severe attacks of headache, commonly affecting only one side of the head.

Malaria: a disease characterized by periodic recurrence of chills, fever and sweating caused by the species of **Plasmodium (e.g., P. vivax or P. falciparum or P. malariae or P. ovale).**

Mania: violent madness.

Measles: also rubeola, acute, highly contagious, fever producing disease caused by a filterable virus, different from the virus that causes the less serious disease *viz.,* German measles, or rubella. Measles is characterized by small red dots appearing on the surface of the skin, irritation of the eyes (especially on exposure to light), coughing, and a runny nose.

Melaena: melanorrhea; melanorhhagia; the passage of dark colored, tarry stools, due to the presence of blood altered by the intestinal juices.

Menorrhogia: excessive bleeding during menstruation.

Micturition: passing urine with blood or sometimes blood alone. **Myasthenia gravis:** chronic disease marked by progressive weakness and abnormally rapid fatigue of the voluntary muscles and characterized by drooping eyelids.

Narcotic: an agent capable of causing a depression in the central nervous system.

Nasal congestion: congestion in the nose.

Nausea: a feeling that vomiting is about to begin.

Nephritic colic: severe pain in the kidney.

Nephritic oedema: swelling or inflammation of kidney.

Nephritis: general term for inflammatory diseases of kidney. **Neuralgia:** nerve pain.

Noctural enuresis: uncontrolled urination at sleep; bed wetting. **Oedema:** edema; an accumulation of an excessive amount of fluid in cells, tissues, or serous cavities.

Oliguria: oliguresia; scanty urination.

Ophthalmia: inflammation of the eye.

Orchitis: inflammation of the testis.

Paralysis: a disease wherein there is loss of power of voluntary movement in any part of the body.

Parkinson's disease: chronic progressive disease usually of the old people and characterized by having muscular tremors, muscular rigidity and general weakness.

Pharyngitis: inflammation of the mucous membrane of the pharynx. **Phlegm:** mucus; self-restraint; calmness; apathy.

Piles: an inflamed condition of the veins in the region of rectum. **Pneumonia:** serious illness with inflammation of one or both lungs usually caused by the infection of *Mycoplasma pneumoniae*. **Poultice:** a soft, usually heated preparation spread on a cloth and applied to an inflamed sore.

Pulmonary catarrh: simple inflammation of a mucus membrane of lungs.

Purgative: an agent that stimulates peristatic action and bowel evacuation.

Pyogenic infection: infection resulting pus-formation.

Pyorrhoea: a disease marked by purulent discharge from the gum. **Rectal fissure:** cleft made by splitting of rectum.

Rectocele: Proctocele; prolapse or herniation of the rectum.

Refrigerant: a drug that relieves feverishness or produces a feeling of coolness.

Resolvent: Discutient; causing resolution; an agent that arrests an inflammatory process or causes the absorption of a neoplasm.

Retropharyngeal abscess: abscess (i.e., a circumscribed collection of pus) arising, usually, in retropharyngeal lymph nodes at the posterior to pharynx, most commonly in infants.

Rheumatism: a term used for pains in the muscles, joints and certain tissues.

Rhinitis: nasal catarrh; inflammation of the nasal mucus membrane. **Rubifacient:** a counterirritant that produces erythema (redness of the skin) when applied to the skin surface.

Scabies: an itching skin disease caused by a mite.

Scalds: burn with hot liquid or stem

Scrofula: tuberculosis of the lymphatic glands.

Scurvy: a deficiency disease caused by lack of vitamin C in the diet, and is characterized by multiple haemorrhage, especially of the gums, gastro-intestinal disturbances and loss of teeth.

Sedative: a drug that reduces excitement, irritation and pain (or has a soothing or sleep-inducing action).

Shrishadi: decoction of the bark of **Albizia lebbeck** used in Ayurvedic treatment.

Shirisharishta: an Ayurvedic preparation from the bark or seed of **Albizia lebbeck.**

Sore: tender and painful part of the body; hurting when touched or used.
Spasm: an intense, uncontrolled muscular contraction.

Spermatorrhea: an involuntary discharge of semen, without orgasm.

Sprain: injure a joint (e.g., wrist or ankle) by twisting violently resulting pain and swelling.

Sputum: matter coughed up from the throat (as indicating the nature of an illness).

Stomachic: a drug that stimulates the appetite, thereby promoting the functional activity of the stomach.

Sudorific: diaphoretic; sudoriferous; sudoriparous; causing perspiration.
Tonic: an agent giving strength and energy.

Tonsillitis: inflammation of the tonsil (two small masses of tissue at the sides of the throat, near the root of the tongue).

Tracheitis: inflammation of the lining membrane of the trachea
Tranquillizer: a drug used to calm or soothe a person without directly inducing sleep.

Trauma: diseased condition of the body produced by a wound or injury; emotional shock, often leading to neurosis (functional derangement caused by disorder of the nervous system).
Triphala: an Ayurvedic preparation comprising fruits of three plants viz., **Terminalia bellirica, Terminalia chebula** and **Emblica officinalis.**

Tubercular lymphadenia: persistent or chronic hyperplasia (i.e., an increase in number of cells in a tissue or organ, excluding tumor) of tubercular nature characterized by the formation of tubercles or small nodules at lymph node.

Ulcer: an open sore (peptic ulcer, ulceration of the stomach, oesophagus or duodenum by the action of the acid in gastric juice).

Urticaria: a vascular reaction of the skin characterized by the transient appearance of irregular elevated patches which are much red or paler than the surrounding skin and often attended by itching.

Uveitis: inflammation of the entire uveal tract: iris, ciliary body, and choroid of the eye.

Vaginal trichomoniasis: disease caused by infection in vagina with a species of ***Trichomonas*** or related genera.

Vaginitis: colpitis; elytritis; inflammation of the vagina.

Vermicide: agent that kills worms.

Vesicant: an irritating substance producing blisters on the skin.

Visceral congestion: congestion in the internal organ of the body (especially the intestines).

REFERENCES

ANONYMOUS 1962. The Wealth of India: Raw Materials. Vol. VI (LM). CSIR New Delhi.

ANONYMOUS 1976. The Wealth of India. Raw Materials. Vol. X (Sp-W). CSIR New Delhi.

ANONYMOUS 1989. Medicinal Plants in China. Compiled by The Institute of Chinese Materia Medica, China Academy of Traditional Chinese Medicine. WHO Regional Publications, Western Pacific Series No. 2, Manila.

ANONYMOUS 1993. Medicinal Plants of Nepal. His Majesty's Government of Nepal, Ministry of Forestry and Soil Conservation, Department of Medicinal Plants, Kathmandu. 153 pp, XXXII.

ANONYMOUS 1994. Ayurveda Authentica: Health in Harmony with Nature. CD-ROM, Tata Interactive System (India) and Dabur Research Foundation, Macromedia.

BHATTACHARJEE S K 1998. Handbook of Medicinal Plants. Pointer Publishers, Jaipur (India), 474 pp.

DASH BHAGWAN 1993. Ayurvedic cures for common Diseases. Fourth Paperback Edition, Hind Pocket Books Pvt. Ltd., Delhi, 200 pp.

EVANS W C 1989. Trease and Evans' Pharmacognosy. Thirteenth Edition. ELBS with Bailliere Tindall, London, 832 pp.

GHIMIRE S K, Y C LAMA, A T N GURUNG and Y A THOMAS 2000. Conservation of Plant Resources, Community Development and Training in Applied Ethnobotany at Shay-Phoksundo National Park and its Buffer zone, Dolpa. Final report of a collaborative project of WWF Nepal Program and People & Plants Initiative. WWF Nepal Program, Report Series # 40.

KATTEL L P 2000. Training Manual for Conservation, Promotion and Utilization of Non-timber Forest Products (in Nepali). BSP/ New Era, Environment and Forest Uddyam Program, Banke.

KIRTIKAR K R, B D BASU and I C S AN 1918. Indian Medicinal Plants Volume II (edited, revised, enlarged and mostly rewritten by E. Blatter, J. F. Caius and K. S. Mhaskar 1981. Lalit Mohan Basu, Allahabad, 1592 pp.

KIRTIKAR and BASU in ANONYMOUS 1985. The Wealth of India. Vol. I: A. CSIR, New Delhi.

MALLA S B and P R SHAKYA 1999. Medicinal Plants: 261-297 in Trilok Chandra Majpuria and Rohit Kumar Majpuria (Edt.) Nepal Nature's Paradise. M. Devi, Gwalior, India, 756 pp.

MALLA S B, P R SHAKYA, K R RAJBHANDARI, M N SUBEDI and B L SHRESTHA 1997. Identification Manual for Selected Non-timber Forest Products of Nepal. FRIS Paper No. 9, Forest Resource Information System Project (FRISP) HMG/ FINNIDA, Metsahallitus - Forest and Park Service, 280 pp.

PRESS J R, K K SHRESTHA and D A SUTTON 2000. Annotated Checklist of the Flowering Plants of Nepal. Darwin Initiative, A Joint Project of The Natural History Museum, London and Central Department of Botany, Tribhuvan University, Kathmandu, 430 pp.

SAMBA MURTY A V S S and N S SUBRAHMANYAM 1989. A Text Book of Economic Botany. Wiley Eastern Limited, New Delhi, 875 pp.

SHRESTHA T B 1999. Nepal Country Report on Biological Diversity. Kathmandu: IUCN Nepal. ix, 133 pp.

SHRESTHA T B and R M JOSHI 1996. Rare, Endemic and Endangered Plants of Nepal. WWF Nepal Program, Kathmandu, Nepal, 244 pp.

INDEX

A

Aanp 96
Abies densa 1
Abies spectabilis 1, 170
Abortifacient 36, 72, 83, 160, 167
Abrus precatorius 1
Abscess 5, 13, 20, 29, 75, 101, 124, 227
Abutilon indicum 2
Abutilon indicum var. populifolium 2
Acacia arabica 4
Acacia catechu xxix, 3
Acacia concinna 4
Acacia nilotica 4
Acacia rugata 4
Achyranthes aspera 5, 172
Achyranthes bidentata 6
Aconite 6, 7, 8, 9
Aconitum bisma 6
Aconitum ferox 6, 7, 9
Aconitum ferox var. laciniata 9
Aconitum gammiei 8
Aconitum heterophyllum 8
Aconitum laciniatum 9
Aconitum palmatum 6
Aconitum spicatum 9, 172
Aconogonum rumicifolium 10, 171
Acorin 11
Acorus calamus 11, 171
Acrid 2, 18, 26, 49, 59, 82, 83, 120, 123, 140, 143, 156
Acute common perilla 117
Aegle marmelos 11
Aesandra butyracea 62
Afim 113
Afim swan 113
African marigold 154

Agrimonia eupatorium 12
Agrimonia pilosa var. nepalensis 12
Ahifen 113
Aindri 19
Ainjeru 164
Ajaka 107
Ajapriya 168
Ajayaphal 27
Ajmoda 21
Ajowan 158
Aksoda 81
Albizia julibrissin 13
Albizia lebbeck 13, 221, 228
Alder 16
Aleha 84, 135
Alexipharmic 6, 82, 83, 94
Alkarka 35
Allium aitchisonii 14
Allium carolinianum 14
Allium fasciculatum 15
Allium gageanum 15
Allium violaceum 15
Allium wallichii 15
Alnus nepalensis 16, 173, 1'. 4
Aloe vera 17
Alpamarisha 18
Alstonia scholaris 17
Alterative 30, 217
Amala 119, 120
Amalaki 119, 120
Amaranthus spinosus 18, 174
Ambah 119
Ambastha 46
Amenorrhea 6, 53, 79, 144
Amnesia 217
Amomum subulatum 19, 174
Ampelopsis japonica var. mollis 19
Amra 96

Anaemia 51, 106, 120, 217
Anagallis arvensis 20
Analgesic 7, 10, 38, 218
Ander 135
Androsace rotundifolia 20
Anesthesia 66
Anodyne 38, 217
Anthelmintic 106, 218
Anthraquinone 17
Antifebrile dichroa 59
Antipyretic 2, 7, 10, 47, 51, 103, 141, 156, 158, 169, 223
Antiseptic 26, 32, 54, 57, 63, 104, 113, 137, 161
Antispasmodic 24, 38, 57, 65, 85, 100, 105, 156, 158
Antitussive 114
Apamarga 5, 6
Aperient 81, 120
Aphonia 84
Aphrodisiac 2, 3, 28, 31, 34, 51, 52, 56, 59, 78, 83, 92, 112, 128, 136, 144, 158, 159, 161, 165, 169
Aphtha 218
Apium graveolens 21
Arachis hypogaea 22
Aragvadh 39
Araka 69
Aralia pseudo-ginseng 112
Arandi 135
Arctium lappa 22
Arisaema intermedium 23
Arisaema tortuosum 23, 175, 176
Arka 35
Artemisia indica 24, 177
Artemisia vulgaris 24
Arum tortuosum 23
Ascariasis 98
Ascaris 219
Ascites 219
Ashoka 139
Ashok tree 139
Asparagus 25, 73
Asparagus recemosus 25, 177

Asthenia 6
Asthma 1, 24, 32, 36, 52, 57, 59, 66, 68, 74, 83, 84, 85, 87, 90, 98, 101, 104, 120, 122, 136, 143, 148, 153, 156, 160, 220
Astilbe rivularis 25, 178, 179
Astringent 1, 2, 5, 6, 12, 26, 27, 30, 31, 33, 34, 41, 44, 55, 56, 66, 71, 82, 93, 97, 98, 100, 109, 112, 137, 140, 145, 153, 156, 160, 162, 166
Asuro 84
Atibala 2
Ativisha 8
Azadirachta indica 26

B

Babbur 4
Babul 4
Badam 22
Badara 168
Badyanchoh 145
Bahira 156
Bakena 98
Bala 145
Balah 156
Balchhar 105
Baliospermum axillare 27
Baliospermum montanum 27
Balu 145
Bankhirro 76
Ban bhendo 38
Ban haloo 53
Ban lasoon 15, 87
Ban lunde 18
Ban tulasi 117
Baran 22
Barbed skullcap 141
Barro 156
Bäsi 127
Bastard myrobalan 156
Batule pat 46
Bauhinia candida 28
Bauhinia purpurea 27
Bauhinia racemosa 28

Bauhinia vahlii 28
Bauhinia variegata 28, 180
Bayar 168
Bayeli 168
Bazar bhang 77
Bead tree 98
Belamcanda chinensis 29
Belleric myrobalan 156
Belu 134
Bengal quince 11
Benzoin neesianum 88
Berberis aristata 29
Berberis asiatica 30, 181
Berberis nepalensis 94
Berberry 29, 30
Bergenia ciliata 182
Bergenia ciliata forma ligulata 31
Bergenia ligulata 31
Bergera koenigii 103
Bermuda grass 54
Besha 53
Bethe 43
Betula bhojpattra 31
Betula utilis 31, 183
Bhalah 143
Bhale-sunpati 134
Bhallataka 143
Bhang 77
Bhanga 37
Bhangerijhar 63
Bhantaki 147
Bhazadri 142
Bhinlaye 63
Bhojpatra 31
Bhorlo 28
Bhote jeera 39, 75
Bhote khair 157
Bhuin amala 119
Bhujinsin 50
Bhumyalaki 119, 120
Bhurjapatra 31
Bhutkesh 142
Bhutle 105
Bhyakur 61
Bibhitaka 156

Bijapura 47
Bikh 7, 9
Bikhma 6
Bilouni 93
Bilva 11
Bisfez 125
Blackberry lily 29
Black creeper 78
Black juniper 82
Black muscale 52
Black plum 153
Blue water lily 107
Boerrhavia diffusa 32
Boerrhavia repens 32
Bojho 11
Bombax ceiba 32
Bombax malabaricum 32
Bosin 16
Box myrtle 104
Brachycorythis obcordata 33
Bromoergocrystine 48
bronchitis 1, 3, 11, 24, 45, 51, 69, 82,
   83, 85, 87, 90, 95, 104, 108,
   119, 120, 122, 133, 136, 153,
   154, 156
Bronchodilator 219
Budhna 115, 116, 130, 162
Budho okhati 25
Butea buteiformis 33, 183, 184
Butea frondosa 34
Butea minor 33
Butea monosperma 34
butylphalide-3 21
Byaha 11

C

Callicarpa incana 35
Callicarpa macrophylla 35
Calotropis procera 35
Caltha bisma 6
Caltha himalensis 36
Caltha palustris var. himalensis 36
Calthrops 159
Camphor 44, 45
Cancer 48, 118, 156, 223

Cannabis sativa 37, 184
Canna edulis 37
Capers 38
Capparis spinosa 38
Caraway 39
Carbuncles 6, 154
Carminative 1, 11, 32, 39, 46, 47, 53, 54, 65, 68, 71, 82, 83, 88, 95, 99, 102, 104, 122, 148, 154, 159, 160
Carotene 43
Carum carvi 39
Carvone 39
Cassia fistula 39, 185
Cassia rhombifolia 39
Cassia tora 40
Castor bean 135
Catechin 139
Cathartic acid 40
Cedrus deodara 41
Cedrus libani var. deodara 41
Celery 21
Celosia argentea 41
Celtis australis 42
Centella asiatica 42
Cephalagia 100
Chakramarda 40
Chalacas koenigii 103
Chamaesyce hirta 66
Chariamilo 111
Charila 115, 116, 130, 162
Chaswan 100
Chaunle 10
Chebulic myrobalan 157
Chenopodium album 43
Cherokee rose 136
Chhatiwan 17
Chhatiwansin 17
Chile kath 131
China berry 98
Chinese date 168
Chinese lobelia 89
Chiraito 149, 150, 151, 152
Chiretta 148, 150, 151, 152
Chitraka 123

Chitraparni 161
Chitu 123
cholagogue 36, 103, 125
Cholera 38, 158
Chopchini 147
Chutro 29, 30
Chyavanprasha 220
Chyuri 62
Cimicifuga foetida 44
Cinnamomum camphora 44
Cinnamomum cassia 45
Cinnamomum glaucescens xxix, 45, 186
Cinnamomum tamala 45, 186
Cirrhosis 81, 90
Cissampelos pareira 46
Citrus limon 47
Claviceps purpurea 47
Clematis 48
Clematis cirrhosa var. nepalensis 48
Clotbur 22
Cluster mallow 96
Cocklebur 22
Coelogyne cristata 49
Colic 6, 11, 18, 21, 24, 34, 39, 46, 56, 66, 68, 93, 95, 100, 130, 140, 146, 158, 226
Colitis 220
Common club moss 92
Common matrimony vine 91
Conception 220
Conessi 76
conjunctivitis 40, 42, 53, 70, 166, 218
Constipation 220
Convalescence 220
convulsion 24, 105, 218, 221
Conyza cappa 79
Coomb teak 72
Cordyceps sinensis xxix, 49
Coriaria nepalensis 50
corneal ulcer 40
Costus nipalensis 50
Costus speciosus 50
Country fig tree 67
Country mallow 145

Crateva religiosa 51
Crateva unilocularis 51
Cryptolepis buchananii 51, 52
Cryptolepis reticulata 52
Curculigo orchioides 52
Curcuma angustifolia 53
Curcuma aromatica 53
Curcuma longa 14, 53
Curry leaf tree 103
Cutch tree 3
Cynodon dactylon 54
Cyperus rotundus 55
Cystitis 75

D

d-selenene 21
Dactylorhiza hatagirea xxix, 55, 187
Dadamari 40
Dahikamlo 35
Dalchini 45
Dale chuk 75
Daphne bholua 56, 188
Daphne cannabina 56
Darim 129
Daruharidra 29
Darya ken 85
Dashang 14, 221
Dasmula 12, 221
Datiwan 5, 6
Datura metel 56, 189
Datura stramonium 57, 189, 190
decoction 3, 14, 16, 29, 30, 34, 36, 47, 67, 73, 92, 104, 107, 130, 137, 144, 161, 165, 228
Delirium 221
Delphinium himalayai 58
Delphinium himalayense 58
Deltoid yam 61
Demulcent 2, 3, 31, 51, 52, 56, 73, 95, 140, 146
Dendrobium nobile 58
Deobstruent 24, 30, 63, 81, 137
depurative 34, 51
Desmostachya bipinnata 59
Dhainyaro 166

Dhasingare 70
Dhataki 166
Dhattura 57
Dhatura 56
Dhukure 69
Dhupi 82, 83
Dhustura 56
diabetes 54, 102, 114, 153
Diamine 18
diarrhea 3, 4, 7, 11, 12, 16, 24, 27, 30, 31, 33, 34, 38, 41, 44, 45, 46, 47, 52, 58, 76, 78, 82, 83, 90, 92, 99, 104, 107, 109, 111, 112, 114, 120, 124, 126, 129, 137, 150, 153, 158, 167, 169, 220, 222
Dichroa febrifuga 59
Dihydroergotoxine 48
Dioscorea bulbifera 60, 190
Dioscorea deltoidea xxviii, 61
Dioscorea nepalensis 61
Dioscorea prazeri 61
Dioscorea sativa 60
Dioscorea sikkimensis 61
Diosgenin 61, 62
Diploknema butyracea 62
Dipya 77
Discutient 227
Dita bark 17
Diuretic 6, 18, 32, 87, 222
Dolichos biflorus 93
Downy datura 56
Dronapushpa 86
Drooping juniper 83
dropsy 5, 32, 36, 77, 156
Drumstick tree 102
Dryopteris filix-mas 62
Dubo 54
Dudela 73
Dudhale 56
Duke of Argylls tea-tree 91
Dumri 67
Durva 54
Dysentery 3, 5, 6, 8, 10, 11, 12, 13, 27, 38, 41, 55, 66, 76, 95, 100,

101, 104, 109, 111, 120, 122,
129, 133, 134, 136, 140, 143,
144, 153, 160, 166, 167, 169
Dyspepsia 6, 9, 11, 12, 51, 107, 120,
140, 150, 158, 217
Dyspnea 222
Dystocia 222

E

Eclipta alba 63
Eclipta prostrata 63
Eczema 111
Eklebir 90
Elaeocarpus sphaericus 64, 191
Elaerocarpus sphaericus 192
Electuary 222
Elephant-grass 160
Ellagic acid 120, 156
Embelia robusta 64
Embelia tsjeriam-cottam 64
Emblica officinalis 119, 228
emetic 5, 6, 11, 36, 38, 63, 65, 110,
112, 128, 139, 160, 167
Emmenagogue 17, 55, 83, 98, 144,
154, 156
Emollient 37, 62, 107, 135, 136, 140,
144, 146
Emphysema 222
Entada phaseoloides 65
Entada scandens 65
Ephedra girardiana 65, 193
Ephedrine 66, 146
Epilepsy 66, 112, 115, 116, 117, 122,
130, 139, 162
Epistaxis 13, 60, 70
Eragrostis cynosuroides 59
Eranda 135
Ergonovine 48
ergot 48
Ervatamia divaricate 153
Euphorbia hirta 66
Everniastrum vexans 115
expectorant 1, 5, 20, 33, 50, 56, 80,
85, 90, 122, 139, 148, 163

F

Faran 15
Feather cockscomb 41
febrifuge 56, 68, 76, 97, 100, 146,
149, 150, 165
Fennel 68
Ficus glomerata 67
Ficus racemosa 67
Fire-flame bush 166
Flemingia strobilifera 67
Foeniculum vulgare 68
Fritillaria cirrhosa 69, 193
Fritillary 69
Fumaria indica 69
Fumaria vaillantii var. indica 69

G

Gaji 37
Gambhari 72
Gamdol 33
Gandhaphali 35
Gandhmadini 164
Ganitrus sphaericus 64
Ganja 37
Gardenia augusta 70
Gardenia jasminoides 70
gastralgia 106
Gastrorrhagia 223
Gaultheria fragrans 70
Gaultheria fragrantissima 70
Gentiana chirayita 150
Geranium nepalense 71, 194
Geranium quinquenerve 71
Ghiukumari 17
Ghodtapre 42
Glaucoma 223
Gloriosa doniana 72
Gloriosa superba 72
Gmelina arborea 72
Gmelina rheedii 72
Gobre sallo 1
Gogan 140
Goitre 223
Gojapu swan 86

Gokhur 159
Gokshura 159
Golden champa 100
Gonorrhoea 4, 32, 145, 159, 160
Gooseberry 119
Gout 223
Greater cardamom 19
Great burdock 22
Guchhi chyau 102
Guduchi 157
Gulancha tinospora 157
Gunja 1
Gunjhamsi 153
Gun labha 87
Gurjo 157
Gya tik 150
Gya tsi goepa 44

H

haematemesis 13, 70
haematuria 13, 70, 85
Haemoptysis 223
Haemorrhage 223
Hairy agrimony 12
Haledo 53
haloo 53
Hamo 143
Harchu 164
Harchur 164
haridra 53
Haritaki 157
Harra 157
Harro 157
Hathan 138
Heal-all 126
Hedera helix 73
Hedera nepalensis 73, 194
Hedychium album 73
Hedychium spicatum 73
Hemdugdhak 67
Hemerocallis disticha 74
Hemerocallis fulva 74
Hemp 37
Henbane 77
hepatitis 70, 96, 155

Heracleum napalense 195
Heracleum nepalense 75
Heracleum nepalense var. bivittata 75
Himalayan silver fir 1
Himalayan wild cherry 127
Hippophae rhamnoides 75
Hippophae tebatana 195
Hippophae tibetana 75
hoarseness 156
Holarrhena antidisentrica 76
Holarrhena pubescens 76
Horsegram 93
Hygrophila auriculata 77
Hygrophila spinosa 77
Hyoscyamus agrestis 77
Hyoscyamus niger var. agrestis 77
hypertension 6, 40, 42, 48, 93, 126, 131
Hypochondrium 224
Hypotension 224
Hysteria 224

I

Ichnocarpus frutescens 78
icteric 70, 96
Imoo 158
Impatiens balsamina 78
Impatiens coccinea 78
impotence 159
Indian blackberry 153
Indian butter tree 62
Indian cassia 45
Indian kudzu 128
Indian laburnum 39
Indian lilac 98
Indian pennywort 42
Indian privet 165
Indian sorrel 111
Indian valerian 162
Indrayani 159
Infusion 32, 93, 149, 224
Insanity 224
insomnia 13, 70, 112, 122
intoxicant 38
Inula cappa 79

Inula eriophora 79
Inula racemosa 79
Inula royleana 79
Ipomoea hederaceus 80
Ipomoea nil 80
Iris decora 81
Iris nepalensis 81
Ironwood tree 99

J

Jamaica wild liquorice 1
Jamane mandro 94
Jambu 153
Jamun 153
Jangali haledo 53
Japanese ampelopsis 19
Japanese honeysuckle 90
Jaringo 121
jatamansi 105
jaundice 21, 30, 32, 52, 59, 70, 75, 76, 87, 95, 106, 120, 224
Java plum 153
Jeewan buti 49
Jhyau xxix, 115, 116, 130, 162
Jimson weed 57
Jiwanti 110
Juglans kamaonia 81
Juglans regia 196, 197
Juglans regia var. kamaonia 81
juniper 82
Juniperus communis var. saxatillis 83
Juniperus excelsa 83
Juniperus indica 82, 196
Juniperus macropoda 83
Juniperus recurva 83
Juniperus sibirica 83
Juniperus wallichiana 82
Justica adhatoda 84
Jwano 158

K

Kagati 47
Kagatpate 56
Kagchalo 65
Kagiyo 2
Kakad singhi 122
Kakoli 69
Kalapanga 69
Kalchelaharo 69
Kalo bikh 9
Kalo dhatura 56
Kalo shariva 78
Kamala tree 95
Kampilla 95
Kampillaka 95
Kanbakan 18
Kancanara 28
Kande loti 155
Kantakadruma 32
Kantakari 148
Kanthaparna 142
Kaphal 104
Kapoo 44
Kapurakachari 73
Karavalli 101
Karingi 166
Karkandhu 168
Karpur 44
Karpura 44
Kasa 138
Katbasi 88
Kemuka 50
Keratitis 224
Keri 87
Khadira 3
Khaibasi 98
Khai kakacha 101
Khalu 148, 150, 151, 152
Khamari 72
Khayer xxix, 3
Khiraule 87
Khirro 166
Kholcha ghayen 42
Khosin 81
Khurpani 126
Kiratatikta 150
Koiralo 27, 28
Kolomba 70
Kovidara 28

Koylikaswan 69
Krishnasariva 52
Kuchrung 157
Kunahbun 28
Kunuh 17
Kurilo 25
Kusalapte 28
Kush 59
Kutaja 76
Kutki 105
Kuvala 168
Kwakhachola 69
Kyaga 85

L

Laghupatra 124
Laincha 78
Lajja 101
Lajjawati 101
Lajwanti 101
Lakshmana 148
Lali gurans 133
Lal gedi 1
Langali 72
Langthang 77
Laptema 144
Laryngitis 224
Laxative 224
lemon 22, 89
Leonurus japonicus 85
Leonurus sibiricus 85
Lepidium apetalum 85
Lepidium ruderale 85
leprosy 2, 26, 36, 43, 54, 55, 72, 119, 122, 129, 136
Leucas capitata 86
Leucas cephalotes 86
Leucoderma 225
Leucorrhoea 225
Lhuchi 62
Lichen xxix, 115, 116, 117, 130, 162
Ligustrum bracteolatum 87
Ligustrum nepalense 87
Lilium nepalense 87
Lilium ochroleucum 87

Lindera neesiana 88
lithiasis 96
Litsea citrata 89
Litsea cubeba 89
Lobelia 89
Lobelia nicotianaefolia 90
Lobelia pyramidalis 90, 197
Lobelia radicans 89
Lodh xxix, 152, 155
Lodhra 152
Lodh salla 155
long pepper 7, 165
Lonicera japonica 90
lumbago 51, 136
Lycium barbarum 91
Lycium halimifolium 91
Lycopodium clavatum 92

M

Machino 70
Macrotyloma uniflorum 93
Madanam 167
Madhraphala 168
Madhuka butyracea 62
Madhurika 68
Maesa chisia 93
Magnolia 100
Mahabala 145
Mahanimba 98
Maharanga emodi 94
Maharangi 94
Mahonia 94
Mahonia napaulensis 198
Mainphal 167
Majino 96
Majitho 137
Malabar glory lily 72
malaria 60, 112, 140
Male fern 62
Mallotus philippinensis 95
Malva alchemillaefolia 96
Malva vercillata 96
Mamira 117
Mandukaparni 42
Manduparni 155

Mangifera indica 96, 198
Mango 96
Mania 225
Manjistha 137
Maripyasi 29, 30
Marking nut tree 143
Marsh marigold 36
Masta 55
measles 23, 160, 225
Megacarpaea polyandra 97
Melaena 225
Meme gudruk 124
Menorrhogia 225
Mentha crispa 98
Mentha spicata 98
Mesua ferrea 99
Mesua nagassarium 99
Mhalavata 28
Mhata 28
Michelia aurantiaca 100
Michelia champaca 100
micturition 75, 137, 159
Mimosa arabica 4
Mimosa catechu 3
Mimosa lebbeck 13
Mimosa pudica 101
Mimosa rugata 4
Mistletoe 164
Momordica charantia 101
Momordica muricata 101
Morchella esculenta 102
Morel mushroom 102
Moringa oleifers 102
Moringa pterygosperma 102
Mothe 55
Mountain ebony 28
Mug-wort 24
Murjhang 121
Murraya koenigii 103, 199
Murva 28
Musabar 17
Musali 52
Mushali 52
myasthenia gravis 66
Mycoplasma pneumoniae 226

Myrica esculenta 104
Myrica farquhariana 104
Myrobalanus chebula 157

N

Nafo 75
Nagbeli 92
Nagkesar 99
Nardostachys gracilis 105
Nardostachys grandiflora xxix, 105
nasal congestion 66
Naswan 105
nausea 90, 118
Nawa ghayen 98
Necator 217
Neem 26
Neopicrorhiza scrophularifolia xxix, 105
Nepalese lily 87
Nepal aconite 9
nephritic colic 21
nephritic oedema 85, 90
nephritis 96
Nerium coronarium 153
Nerium divaricatum 153
Nettle tree 42
neuralgia 19, 78, 99
Nhayekan 161
Night jasmine 106
Nilkamal 107
Nimba 26
Nirbisi 117
Nirgundi 165
Nirmashi 8
Nisoth 108
Noble dendrobium 58
noctural enuresis 66
Nun dhiki 110
Nut grass 55
Nyctanthus arbor-tristis 106
Nymphaea stellata 107

O

Ocimum sanctum 107

Ocimum tenuiflorum 107, 199
Okhar 81
oliguria 85
On 96
Operculina turpethum 108
Ophelia bimaculata 149
Ophthalmia 226
Opium poppy 113
Orchis hatagirea 55
orchitis 60
Oroxylum indicum 108
Osbeckia nepalensis 109
Osyris arborea 110
Osyris wightiana 110
Otochillus porrectus 110
Oxalis corniculata 111
Oxalis pusilla 111

P

Padamchal 132
Padbiri 111
Padmaka 127
Paederia foetida 111
Paiyun 127
Palabi 33
Palija swan 106
Panax pseudo-ginseng 112
Panchaunle 55
Pandanus diodon 112
Pandanus nepalensis 112
Pangra 65
Panicum dactylon 54
Panlas 34
Papaver amoenum 113
Papaver sominiferum 113
Papra 124
Paralysis 226
Parijata 106
Paris marmorata 114
Paris polyphylla 200
Paris polyphylla subsp. marmorata 114
Parmelia cirrhata 115
Parmelia nepalensis 115
Parmelia nilgherrensis 116

Parmelia tinctorium 116
Parmotrema nilgherrense 116
Parmotrema tinctorum 116
Parnassia nubicola 117, 201
Pashanved 31
Patana 98
Patha 46
Patpate 70
Paunghayen 111
Peach 127
Peanut 22
Peppergrass 85
Pepperweed 85
Perilla frutescens 117
Perilla ocimoides 117
Persian lilac 98
Phalisi 95
Pharbitis nil 80
pharyngitis 60, 84
Phenila 138
phlegm 223
Pholidota articulata 118
Phyllanthus amarus 119
Phyllanthus emblica 119, 201, 220
Phyllanthus niruri 119
Phyllanthus taxifolius 119
Phyllanthus urinaria 120
Phytolacca acinosa 121
Phytolacca latbenia 121
Picrorhiza kurroa 105
Picrorhiza scrophulariflora 105
Pigweed 43
piles 3, 5, 6, 17, 34, 52, 65, 82, 83, 93, 94, 95, 101, 107, 112, 124, 140, 144, 145, 156
Pipala 121
Pipalamul 121
Pipee 121
Piperine 122
Piper longum 7, 14, 121, 165
Pippali 121
Pistacea chinensis 122
Plectranthus cordifolia 123
Plectranthus incanus 123
Plectranthus mollis 123

Plumbago rosea 123
Plumbago zeylanica 123
Pneumonia 226
Podinak 98
Podophyllin 124
Podophyllum emodi 124
Podophyllum hexandrum xxviii, 124, 202
Polyacetylenic thiophenes 63
Polypodium vulgare 125
Pomgranate 129
Pootiha 98
Potentilla fulgens var. intermedia 125
Potentilla josephiana 125
Potentilla lineata var. intermedia 125
Potentilla microphylla 202
Prapunnada 39, 40
Prasarini 111
Prickly amaranth 18
Prickly chaff 6
Pride of India 98
Prishniparni 161
Priyangu 35
Prunella vulgaris 126
Prunus armeniaca 126
Prunus creasoides 127
Prunus persica 127
Prunus puddum 127
pudina 98
Pueraria tuberosa 128
Pukarmula 79
pulmonary catarrh 84
Punarnava 32
Punica granatum 203
Punica granetum 129
purgative 2, 5, 17, 24, 27, 40, 56, 63, 68, 82, 95, 108, 109, 124, 128, 135, 136, 160, 219, 224
Purging cassia 39
Puskara 79
Puskara mula 79
pyogenic infection 20, 85

Q

Queensland arrowroot 37
Quisqualis indica 129

R

Raasna 79
Ragwort 142
Rajbriksha 39
Ramalina 130
Randia spinosa 167
Rangoon creeper 129
Rasanjan 29, 30
Rasula 4
Rati gedi 1
Rato lahare ghans 66
Rauvolfia serpentina xxviii, xxix, 131, 203
Rectal fissure 227
Rectocele 227
Refrigerant 227
Resolvent 227
Retropharyngeal abscess 227
Rhamnus napalensis 131
Rheumatism 227
Rheum australe 132, 204
Rheum emodi 132
Rheum nobile 132
Rhinitis 227
Rhododendron 133, 134, 204, 206
Rhododendron arboreum 133, 205
Rhododendron lepidotum 134, 206
Rhododendron puniceum 133
Rhynchostylis retusa 135
Riccinus communis 207
Ricinus communis 135
Ritha 138
Rochani 98
Rohini 95
Rosa laevigata 136, 206, 207
Rosa sinica 136
Rottlerin 95
Rubia cordifolia 137
Rubia manjith 73, 137
Rudraksha 64

Ruga sag 97

S

Saccharum canaliculatum 138
Saccharum spontaneum 138
Sala 144
Sallejari 65
Sano chilya 145
Sansuh 95
Säpi 11
Sapindus mukorossi 138
Sapphire berry 152
Saptaparna 17
Saraca asoca 139
Saraca indica 139
Sarpagandha xxix, 131
Sarpa Makai 23
Satawari 25
Satuwa 114
Satyrium nepalense 140
Saurauia napaulensis 140
Saurauia paniculata 140
Sayepatri 154
scabies 159
Scalds 227
Schima wallichii 141
Screw pine 112
Scrofula 227
Scurvy 227
Scutellaria barbata 141
Scutellaria peregrina 141
Selenium wallichianum 207
Selfheal 126
Selinum tenuifolium 142
Selinum wallichianum 142
Semecarpus anacardium 143
Sensitive plant 101
Sephalika 106
Serpent wood 131
Sesame 143
Sesamum indicum 143
Sesamum orientale 143, 208
Seto chulsi 109
Shalparni 67
Shirish 13

Shirishadi 14
Shirisharishta 228
Shobhanjana 102
Shorea robusta xxix, 144, 209
Shyonaka 108
Sicklewort 126
Sickle senna 40
Sida compressa 145
Sida cordifolia 145
Sida rhombifolia 145
Sigesbeckia brachiata 146
Sigesbeckia orientalis 146
Sigru 102
Sikakai 4
Silk tree 13
Silpigan 51
Siltimur 88, 89
Simal 32
Simali 165
Sindure 95
Sirisha 13
Sisnu 161
Sitavarka 41
Skunk bugbane 44
Smaller morning glory 80
Smilax aspera 147
Smilax capitata 147
Soapnut tree 138
Solanum anguivi 147
Solanum diffusum 148
Solanum indicum 147
Soma 65
Somlata 65
Soochyagra 59
sore 6, 15, 23, 29, 44, 54, 148, 153, 161, 221, 226, 229
Spearmint 98
Spiked ginger lily 73
Spiny amaranth 18
Spiral ginger 50
Sponge mushroom 102
Sprain 228
Sthulaela 19
Sthulapushpa 154
Stinging-nettle 161

Stomachic 17, 228
Sudorific 228
Sugandhakokila xxix, 45
Sugandhawal 162
Sunpati 133
Surabhi 79
Sushavi 39, 101
Suvarnaka 39
Swallow-wort 35
Swertia alata 148
Swertia angustifolia 149
Swertia angustifolia var. hamiltoniana 149
Swertia bimaculata 149
Swertia bimaculata var. macrocarpa 149
Swertia chirata 150
Swertia chirayita 150, 210
Swertia ciliata 151
Swertia dilatata 152
Swertia multicaulis 151
Swertia paniculata 152
Swertia purpurascens 151
Symplocos chinensis 152
Symplocos paniculata 152
Syzygium cumini 153, 211

T

Tabernaemontana coronaria 153
Tabernaemontana divaricata 153
Tagar 153
Tagetes erecta 154
Takswan 154
Talamuli 52
Talisa 155
Talispatra xxix, 1, 155
Talmakhan 77
Tamalaka 45
Tandula 18
Tanki 27
tannin 5, 12, 29, 40, 44, 101, 120, 123, 126, 129, 156
Taphoswan 154
Taraxacum officinale 155
Taraxacum officinale var. parvulum 155
Tarika 112
Tarul 60
Tatelo 108
Taxus baccata 155
Taxus wallichiana xxix, 155, 211, 212
Ta mik chewa 36
Tejpat 45
Terminalia bellirica 156, 228
Terminalia chebula 157, 228
Tetrapeltis fragrans 110
Thorn apple 56
Thulo okhati 25
tiger lily 29
Tikta 150
Til 143
Timoo 167
Timur 167
Tinospora cordifolia 157
Tinospora sinensis 157
Tite 101, 149, 151, 152
Titepati 24
Tite karela 101
tonic 1, 5, 7, 8, 11, 26, 28, 36, 38, 43, 47, 50, 51, 52, 53, 54, 56, 68, 76, 78, 80, 83, 93, 97, 102, 103, 105, 109, 110, 112, 114, 120, 128, 136, 137, 140, 144, 145, 149, 150, 155, 158, 165, 169
Tonsillitis 228
tracheitis 29
Trachyspermum ammi 158
Trauma 228
Tribrit 108
Tribulus terrestris 159
Trichomonas 229
Trichosanthes bracteata 159
Trichosanthes tricuspidata 159
Trichosanthus 212, 213
Triphala 120, 156, 228
Tubercular lymphadenia 229
Tukiphool 155
Tulasi 107
Tulsi 107
Tumburu 167

Turmeric 53, 54
Turpeth root 108
Typha angustifolia 160
Typha elephantina 160
Tyuri 78

U

Udumber 67
ulcer 23, 35, 40, 67, 70, 124, 153, 157, 218, 221, 229
Umbrella tree 98
Umma lauri 157
Unyu 62
Uraria picta 161
Urista 138
Urticaria 229
Urtica dioica 161, 213
Usnea thomsonii 162
Utis 16
Utrasum bead tree 64
Uveitis 229

V

Vacha 11
vaginal trichomoniasis 13
vaginitis 76
Valeriana jatamansii 162, 214
Valeriana spica 162
Valeriana wallichii xxix
Vana-haridra 53
Vanda alpina 163
Vanda cristata 163
Varahi 60
Vasaca 84
Vasaka 84
Vasicine 85
Vayuvidanga 64
Vermicide 229
Vesicant 229
Vidari 128
Vijaya 37
visceral congestion 155
Viscum album 164
Viscum articulatum 164

Viscum costatum 164
Viscum nepalense 164
Vitex negundo 165
Vitunna 41
Vrihati 147

W

Wall fern 125
Walnut 81
Water pennywort 42
Wax myrtle 104
White flower leadwort 123
Wild turmeric 53
Wintergreen 70
Woodfordia floribunda 166
Woodfordia fructicosa 166
Wormwood-like motherwort 85
Wrightia arborea 166
Wrightia mollissima 166
Wrightia tomentosa 166

X

Xeromphis spinosa 167

Y

Yagyabhooshan 59
yam 60, 61
Yarsagumba 49
Yavani 158
Yellow champa 100
Yerba de tajo 63
Yew 155

Z

Zanthoxylum alatum 167
Zanthoxylum armatum 167, 214, 215
Ziziphus jujuba 168
Ziziphus mauritiana 168

About The Authors

Kamal Krishna Joshi Ph.D. is former Vice-chancellor of Tribhuvan University, Kathmandu, Nepal. He taught Genetics and Economic Botany to the postgraduate students in the Central Department of Botany of Tribhuvan University for 22 years. He is the author of about 60 research papers and popular articles published in local and international journals, magazines and other periodicals. He has already published two books; one of them is "Samajik Anuvanshiki" (Social Genetics) in Nepali and the other is "Genetic Heritage of Medicinal and Aromatic Plants of Nepal Himalayas". The latter was written in joint authorship with his wife Prof. Dr. Sanu Devi Joshi. He traveled throughout Nepal including most of the Himalayan ranges.

Dr. Joshi delivered invited lectures and keynote addresses on different occasions held in Belarussian State Polytechnic Academy, Minsk, Belarus; Josai International University, Tokyo; Lakehead University, Canada; Prajapita Bramhakumari Ishwariya Vishwavidyalaya, Mount Abu, Rajasthan, India; DAAD Seminar for Former Grantees, Kathmandu, Nepal and address to the Senate members of the Sankt Petersburg State Mining Institute (Technical University), Sankt Petersburg, Russia.

He is decorated with many national and international awards including Prashidda Praval Gorakha Dakshin Bahu (Go.da.ba.II) by HM the late King Birendra Bir Bikram Shah Dev of Nepal, Highest Honour of Soka University, Tokyo, Ambassador For Peace designated by the Interreligious and International Federation for World Peace, awards and letter of appreciation from State Science and Technology Commission of People's Republic of China and letter of Honour from Nepal University Teachers' Association. At present, he is the Chairman of Himalayan Botanical Research Center P. Ltd. and Xavier International College, Kathmandu. He pioneered the campaign "Science for the Common Man" in Nepal more than two and half decades back and still engaged in the campaign.

Sanu Devi Joshi Ph.D. is a Professor of Botany in the Central Department of Botany of Tribhuvan University, Kathmandu. She is specialized in Plant Biotechnology and Cytogenetics of liverworts and ferns. There are many research papers and popular articles published in local and international journals, magazines and other periodicals to her credit. She is teaching biosystematics of lower plants and gymnosperms and Biotechnology to postgraduate students for last 36 years in Central Department of Botany, Tribhuvan University, Kathmandu. She has successfully completed several research projects related to ex-situ conservation of threatened and endangered plants of Nepal and genecology of ***Alnus nepalensis*** D. Don sponsored by Gorkha Ayurved Company P. Ltd., International Center for Integrated Mountain Development (ICIMOD) and Royal Nepal Academy of Science and Technology (RONAST) respectively.

Prof. Joshi is decorated with Prabal Gorkha Dakshin Bahu (Go da ba III) from HM the late King Birendra Bir BikramShaha Dev, Highset Honour of Soka Women's College, Tokyo, Ambassador For Peace designated by the Interreligious and International Federation for World Peace, awards and letter of appreciation from State Science and Technology Commission of People's Republic of China.

She has traveled many places of Nepal and abroad in connection with her research projects and participation in conferences and talks etc. She is a co-author of the book entitled "Genetic Heritage of Medicinal and Aromatic Plants of Nepal Himalayas".